T0135966

Proceedings

of the

International Beilstein Symposium

on

SYSTEMS CHEMISTRY

May 26th – 30th, 2008

Bozen, Italy

Edited by Martin G. Hicks and Carsten Kettner

 **Beilstein-Institut**

Systems Chemistry, May 26th – 30th, 2008, Bozen, Italy

BEILSTEIN-INSTITUT ZUR FÖRDERUNG DER CHEMISCHEN WISSENSCHAFTEN

Trakehner Str. 7 – 9
60487 Frankfurt
Germany

| **Telephone:** | +49 (0)69 7167 3211 | **E-Mail:** | info@beilstein-institut.de |
| **Fax:** | +49 (0)69 7167 3219 | **Web-Page:** | www.beilstein-institut.de |

IMPRESSUM

Systems Chemistry, Martin G. Hicks and Carsten Kettner (Eds.), Proceedings of the Beilstein-Institut Symposium, May 26th – 30th 2008, Bozen, Italy.

Bibliographic information published by the *Deutsche Bibliothek*.
The *Deutsche Bibliothek* lists this publication in the *Deutsche Nationalbibliografie*; detailed bibliographic data are available in the Internet at http://dnb.ddb.de.

ISBN 978-3-8325-2188-2

Layout by:	Hübner Electronic Publishing GmbH	Printed by	Logos Verlag Berlin GmbH
	Steinheimer Straße 22a		Comeniushof, Gubener Str. 47
	65343 Eltville		10243 Berlin
Cover Illustration by:	SEIBERT MEDIA GmbH		http://www.logos-verlag.de
	Söhnleinstr. 8		
	65201 Wiesbaden		

PREFACE

The Beilstein symposia address contemporary issues in the chemical and related sciences by employing an interdisciplinary approach. Scientists from a wide range of areas – often outside chemistry – are invited to present aspects of their work for discussion with the aim of not only advancing science, but also, furthering interdisciplinary communication.

Chemistry and biology are two of the most creative sciences. The ability of chemists to design and create their own research objects is a unique feature of this science, bringing it close to art. The aesthetics of symmetry, of biomolecules, or of an elegant synthesis, dissolve the boundaries between art and science. The unique art of biological systems, often unrivalled in the degrees of scale, regularly provides inspiration for chemists and biologists striving for a greater understanding of nature.

Understanding of chemical and biological systems has often been best achieved through reductionism; the bottom-up approach in going from small reaction systems to more complex systems consisting of hundreds or thousands of components is usually impractical. Complex problems are broken down into smaller parts, on the assumption that these behave in predictable, reproducible ways so that new theories or methods can be developed, tested and refined. For example, chemistry has been used very creatively to help understand pharmacological systems. Modern biology through point mutations, siRNA, cloning and knockouts, also provides many creative tools to allow many insights into complex biological systems.

An underlying theme of the symposium was the quest to increase our understanding of nature going from methodologies with regard to chemical building blocks, to complex molecules, supramolecular assemblies, cells and organisms. Complex chemical systems are, of course, not only biological in nature; comprehension of the underlying chemistry, in particular at the nano or meso-scale, of molecular organization allows a systems science approach to be applied to chemistry. Now that biologists and chemists are becoming able to modify and control biological systems, using the combined creativity and prowess of both disciplines, many hidden secrets of the biological systems in cells and organisms can be begun to be understood and investigated in a structured manner. The many parallels between contemporary chemistry and complex biological processes are resulting in innovative research projects throughout the world.

The secluded setting of Hotel Schloss Korb and its convivial atmosphere provided once again the ideal location for the symposium and the ready exchange of thoughts and ideas. Of course, despite the great efforts of all participants, not all scientific problems could be solved over the three days, but many very interesting discussions were initiated which continued well after the symposium; we will be watching the evolution of systems chemistry with much interest over the next years.

We would like to thank particularly the authors who provided us with written versions of the papers that they presented. Special thanks go to all those involved with the preparation and organization of the symposium, to the chairmen who piloted us successfully through the sessions and to the speakers and participants for their contribution in making this symposium a success.

Frankfurt/Main, March 2009   Martin G. Hicks
                                                         Carsten Kettner

CONTENTS

Page

Page

 **Beilstein-Institut** Systems Chemistry, May 26th – 30th, 2008, Bozen, Italy

FRUSTRATION: PHYSICO-CHEMICAL PREREQUISITES FOR THE CONSTRUCTION OF A SYNTHETIC CELL

ANTOINE DANCHIN[*] AND AGNIESZKA SEKOWSKA

Genetics of Bacterial Genomes – CNRS URA2171,
Institut Pasteur, 8 rue du Docteur Roux, 75724 Paris, France

E-Mail: *antoine.danchin@normalesup.org

Received: 7th August 2008 / Published: 16th March 2009

ABSTRACT

To construct a synthetic cell we need to understand the rules that permit life. A central idea in modern biology is that in addition to the four entities making reality, matter, energy, space and time, a fifth one, information, plays a central role. As a consequence of this central importance of the management of information, the bacterial cell is organised as a Turing machine, where the machine, with its compartments defining an inside and an outside and its metabolism, reads and expresses the genetic program carried by the genome. This highly abstract organisation is implemented using concrete objects and dynamics, and this is at the cost of repeated incompatibilities (frustration), which need to be sorted out by appropriate «patches». After describing the organisation of the genome into the paleome (sustaining and propagating life) and the cenome (permitting life in context), we describe some chemical hurdles that the cell as to cope with, ending with the specific case of the methionine salvage pathway.

An Introduction: Three Revolutions and the Birth of Synthetic (Symplectic) Biology

The past century witnessed a remarkable development of biological sciences, split between three overlapping revolutions. The 1944 – 1985 period saw the formation and development of molecular biology, with the creation of all the central concepts of modern biology, combining biochemistry and genetics, and deciphering the rules of replication and of gene expression with its regulation. The advent of gene sequencing technologies in 1975 permitted access to the exact text of genes. Subsequently, the years 1985 – 2005 saw a fascinating development of genomics with the central discovery (in 1991 at a EU meeting in Elounda, in Crete, but published later [1, 2]) that, contrary to expectation, a huge number of genes was of unknown function. At the time of writing (July 22nd, 2008), there is 3,887 ongoing genome sequencing projects, 833 already completed, (mostly from microbes, among which 686 of Bacteria, more or less correctly annotated) and 209,035,780,490 nucleotides are registered at the International Nucleotide Sequence Database Collaboration (INSDC). In this collection, 10% correspond to the core genome ("persistent" genes), while 40 – 50% coding DNA sequences (CDSs) do not correspond to known functions, showing that we lack understanding of a considerable fraction of what makes a living system.

As time elapsed the importance of the relationships between the objects of life – not necessarily the objects themselves – was recognized as absolutely central to any attempt to understand biological processes [3]. Nevertheless, the most recent avatar of molecular biology, Synthetic Biology, was launched initially as an engineering (and teaching) attempt meant to explore the following question:

> "Can simple biological systems be built from standard, interchangeable parts and operated in living cells? Or is biology simply too complicated to be engineered in this way?"
> (iGEM home page: http://parts2.mit.edu/wiki/index.php/About_iGEM).

This technological aim tries to class and normalise "biobricks", basic components of living organisms, while the most important conceptual aim of Synthetic Biology is to reconstruct life, in an endeavour to explore whether we understand what life is and learn missing entities (physical objects and dynamic processes) from our failures. Keeping the abstract laws defining life, a second aim (sometimes named "orthogonal" synthetic biology) tries to reconstruct artificial living systems made from objects of a physico-chemical nature differing from that of the building blocks of extant life [4].

In all these aspects of Synthetic Biology, dynamic processes and rules of interactions appear to be essential, so that we should rather think of this new area of science as *Symplectic Biology* – from the Greek equivalent of the latin *complexus*, meaning 'to weave together' – which would combine the efforts of systems biology with engineering biology [4]. A further (sociological) argument to prefer the latter name is to avoid the fuzzy (and self-contradictory) connotations associated to the word "complexity". Another reason to prefer a term that does not have strong connotations outside biology is that we need to avoid the confusion that plagues the understanding by the general public of the construction of genetically modified organisms. A connotation in geometry ("symplectic geometry" is a lively domain of mathematics) will not interfere. Constructing a synthetic cell requires to understand what life is. In what follows I try to point out features that will be essential to take into account to make a cell *de novo*.

WHAT LIFE IS

Since the time when Schrödinger proposed his famous metaphor of the "aperiodic crystal" major discoveries accumulated that result today in a way to consider life as the association of a machine and of a program. The machine, which expresses the program, is made of a casing that defines an inside and an outside and drives exchanges within and without. It is also a chassis constraining the form of the living organism, with the cell as its "atom".

Compartmentalisation is essential to life and there are two major scenarios associated to this process. Either the cell is made of one single entity, encased in a more or less complex envelope (this corresponds to the domain prokaryotes), or the organism multiplies membranes and skins, even at the cell level, which comprises a nucleus and a variety of organelles (this corresponds to the domain eukaryotes).

The machine also organizes chemical processes – *metabolism* – that build up, salvage and turn over all the required elements making the cell as well as the energy needed to make it work. Metabolic activities are at the root of the reproduction process, which preserves the relationships between pathways, in time and space but not necessarily in their ultimate details. At least 800 small molecules, assimilating C, H, N, O, S, P in the presence of specific ions (note that the role of iron is probably underestimated, as ferrous iron oxidizes extremely rapidly in the presence of dioxygen, and then precipitates in neutral or alkaline water [5]) are involved in the building up of the biomass. Energy is managed via the turnover of ATP and electron transfers. While the number of basic building blocks is small, many investigators tend to forget the importance of co-factors (co-enzymes and prosthetic groups), that are present generally at quite low concentrations but are essential for life. In this respect it is amusing to remark that most studies claiming to work on the origin of life forget about cofactors.

Associated to the machine is a program, involved in processes which may be collectively summarized as *"information transfers"*. It acts as a book of recipes, or, following the common metaphor, as a blueprint. A noteworthy feature of these information transfers is that they are *recursive*, using a code, a cypher, that permits one level of information to be translated into another level, the latter permitting synthesis of objects that can manipulate the program which encoded them. Life is therefore witnessing one of those exceptionally rich "strange loops" (as they were recognised and named by Douglas Hofstadter [6]), which were used by Kurt Gödel, coding axioms and definitions of arithmetic as integers, to demonstrate the incompleteness of arithmetic.

A remarkable feature of this separation between the machine and the program is that it leads one to distinguish between *reproduction* and *replication*. While the latter inevitably accumulates errors [7, 8], the former can improve over time [9]. This is witnessed by the remarkable, but unobtrusive paradox that it is always an aged organism which give birth to young ones [10]. Hence, living organisms have an in-built capacity to generate information.

While the word "information" is currently used in biology, its meaning is not accurately defined [11]. This widespread use nevertheless emphasises the need to add a fifth entity to the four entities considered in classical physics to account for Reality, matter, energy, space and time, which are associated together in the remarkably concise equation proposed by Einstein, $E = mc^2$. While not compatible with classical physics, Heisenberg's indeterminacy principle, $\Delta x \, \Delta p \geq h/4\pi$, introduces information via "lack of information". In a nutshell, I contend that we are at the dawn of a new era in natural sciences, where *information* will play an ever increasing role as we will better understand and model the concept. The core of our future exploration will be to try and understand how information is articulated with matter, energy, space and time. This view implies a considerable change in the placing of biology in the Auguste Comte's hierarchy of sciences, according to increase in information, and progressively less influence of matter, energy, space and time:

$$
\begin{array}{ll}
\text{M/E/S/T} & \quad\quad \Bigg\downarrow \text{information} \\
\\
\text{Classical physics} & \\
\text{Quantum physics} & \\
\text{Chemistry} & \\
\text{Biology} & \\
\text{Neurobiology} & \\
\text{Linguistics} & \\
\text{Mathematics} & \\
\end{array}
$$

With this view, biology is strongly linked to mathematics, and it needs to be perceived essentially as an information-related science. This also indicates that we are in considerable need, at present, to develop further views of what information is. Claude Shannon has investigated the constraints operating on communication of information, not on information itself [3, 11], and many further views have been developed, along a path which is certainly very preliminary but already quite rich conceptually [12, 13].

COMPUTING

In the cell, information transfer is organized by the genetic program. If we take seriously the view just outlined, this process is much more than a metaphor: do we have the conceptual tools to push it to its ultimate consequences? Let us consider what computing is. As demonstrated by Alan Turing and many others [14 – 16], two entities are required to permit computing organised as a *machine* able to read and write a *program* on a physical support. The program is split by the human mind (not conceptually!) into two entities, the program itself (providing the "goal", in our anthropocentric view) and the data (providing the context). An essential point in this description is that the machine is physically distinct from the data/program and can be separated from it. Another point, which is not discussed here is that what we name "program" is *declarative* ("I am here", is enough to start running the program) not *prescriptive*.

Can we see cells as computers, or, asked otherwise, is the genetic program *separated* from the cell's machinery? At least four lines of evidence argue in favour of this view:

- Horizontal gene transfer is extremely widespread. In bacteria, it corresponds often to at least one fifth of the genome setup [17]. This indicates that the cell machinery can "understand" (i.e. read and express) a huge number of genes present in the environment. As a matter of fact, for a given bacterial species (with the caveat that "species" is difficult to define in the case of bacteria), the number of genes that can be horizontally transferred greatly outnumbers the average number of genes present in a given strain. For *Escherichia coli*, for example, taking into account the sequences of published strains, the number of genes that differ from strain to strain is already larger than 20,000, and this number keeps increasing as new strains' genomes are sequenced, while the average number of genes in any strain of this organism is slightly higher than 4,000.

- Viruses behave as pieces of program with a casing allowing them to recognize the machine they will parasite, and a process for coding for their own replication. In this case the metaphor went the other way around: computer scientists will speak about computer viruses, and this is a correct way to describe these invading, often noxious, pieces of programs. As in the case of biological viruses, computer viruses can not only replicate, but they can also carry information loads they extracted from previous infectious cycles. One notes that with this definition a virus is not living (it lacks the machine, and in particular the whole recursive translation machinery, even when it carries genes extracted from a variety of cells and coding for some functions involved in information transfers).

- A further way toward the "computing automaton" view of the cell is the process of genetic engineering. Here, not only do we have cases where genes are artificially associated together, but it is current practice to get DNA sequence pieces that are entirely synthesised from scratch, after purposeful design (this is the only instance of real intelligent design...).

- Finally, the most interesting experiment demonstrating that the program is separated from the machine is the direct transplantation of a naked genome into a recipient cell with subsequent change of the recipient machine into a new one corresponding to the transplanted DNA [18] (Figure 1).

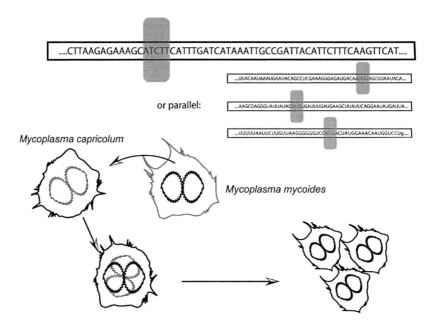

Figure 1. The Turing Machine and an experiment of chromosome transplantation. The Universal Turing Machine head reads and write on a linear string of symbols. Specific Turing Machines can work with parallel pieces of program (here illustrated in the case of protein translation starts). In a transplantation experiment DNA from *Mycoplasma mycoides* is transplanted in *M. capricolum* under selective conditions. The resulting colony is typical of *M. mycoides* [18].

All these observations point to an obvious separation between a "machine" (the cell factory) and "data/program" (the genome). This provides a convincing background to analyse the way information is transferred in living processes.

The Universal Turing Machine works on a program made of one linear string of symbols. Turing has further shown that this is equivalent to a machine with a parallel setup, where several pieces of program could run in parallel. The organisation of information transfer in the cell is more of the latter type, when many pieces of DNA are translated into proteins, for example. Parallel information processing requires coordination, or a clock. In general, biological information transfers are algorithmic in nature. Replication, transcription as well as translation display a high parallelism, always expressed along the same pattern: "Begin, control check-points, repeat, end". The information transfer action is oriented, with a beginning and an end. Curiously, the processes of time dependent control (check-points, or clocks) are rarely taken into account (except for the replication/division processes [19]), but their role is essential to allow the coordination of multiple actions in parallel. This is a first prediction of the model of the cell-as-a-computer: it should prompt investigators to construct experiments to identify check-points in the processes of transcription and translation. Some experiments suggest that they do exist [20, 21].

A MAP OF THE CELL IN THE CHROMOSOME?

John von Neumann, trying to understand the functioning of the brain, suggested that, were a computer both to behave as a computer and to construct the machine itself, it should keep somewhere an image of the machine [16]. The metaphor does not appear to apply to the brain, does it apply to the cell? Linking a geometric program to the information of the genetic program may seem farfetched.

However we have one – unexplained – example of such a link. The homeogenes found in insects follow an order that exactly matches that of the segments of the insect. The comparison between insect and crustacean substantiates this observation: Geoffroy Saint-Hilaire in the middle of the 19th century showed that the body plan of crustacea was reversed under the thorax (the abdomen becomes the back and vice versa) as compared to that of insects, and this triggered a bitter controversy. This has now been proven and backed by the observation that modification of homeogenes between insects and crustacean affects their body plan [22, 23]. The same is true for vertebrates, where four sets of corresponding homeogenes also match the organisation of the adult organism.

We thus have the equivalent of the *homunculus* of preformists, but not as a full tiny organism, but, rather, as the algorithm for the construction of the organism. Can we think of a "celluloculus"? Stated otherwise, is there an image of the cell in the genome? An analysis of the *mur-fts* clusters by Tamames and co-workers suggests that this may be as follows: a tree built up following the way the corresponding genes distribute in different bacterial genomes parallels the bacterial shape variations, not the 16S phylogenetic tree [24]. All this points to the need to explore the points of contact between the information setup and the material setup of living organisms.

FRUSTRATION OF DNA STRUCTURES

The organisation just described is conceptual, it deals with immaterial information, while it needs to be implemented concretely, within the matter/energy/space/time dimensions. However, concrete objects have often properties that are not compatible with those of other objects. This implies "frustration" of possible mutually exclusive entities (because of constraints in space or energy states) [25]. The cell factory will therefore require construction of appropriate "patches" to cope with these incompatibilities.

As a first example, the cell-as-a-computer model requires check-points for parallel gene expression, and this introduces a need for regulation (which may be seen as an important constraint at the origin of the creation of the various regulation systems that are pervasive in biology). This results in a large number of mutually exclusive constraints which are typical of a ubiquitous type of frustration, and explains why it is often so difficult to sort out transcriptional controls mediated by different factors interacting with the same promoter region.

A second physico-chemical constraint derives from the fact that the program needs to be physically separated from the machine. Interestingly, this particular feature matches a common objection raised against the model of the cell-as-acomputer: in living cells, it is not possible to completely separate between the hardware and the software. However, the objection cannot be retained as a strong one, as the same holds true for real computers. Indeed, these machines cannot be purely abstract entities either, but are very concrete entities. They run programs, but any program needs a physical support. For example it can be stored on a CD, and a CD is deformable, by heat for example. When deformed, and despite the fact that the program it carries is unaltered, the laser beam that is used to read it will not be able to do so, and the program will no longer be usable by the computer (Figure 2). This does not alter the very existence of either the computer or the abstract laws establishing what a computer is (a Turing Machine) but this tells us that in any concrete implementation of the Turing Machine, one cannot completely separate between the hardware and the software. This observation points to an important constraint that may explain the somewhat surprising lack of a transplantation experiment in the recent synthesis of an artificial *Mycoplasma* genome [26].

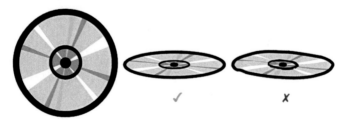

Figure 2. A computer's program must be carried by a physical support. Here, a deformed CD can no longer start a computer.

Still another constraint results from the dissymmetry of replication, conservative on the leading strand and semi-conservative on the lagging strand. This lack of symmetry implies that mutational errors and efficiency of repair will differentially affect the nucleotide and gene composition of both strands of the double helix [27]. As a matter of fact genes that are found to be essential in the laboratory are systematically located in the leading replication strand [28]. This has considerable consequences in the amino acid composition of proteins [29]. In short, material implementation of a Turing Machine requires a variety of specific adjustments to manage material and temporal incompatibilities.

FRUSTRATION IN PROTEINS BUILD-UP

The major effectors of cell metabolism are proteins. Their activity usually requires functional interactions, and it is expected (and observed) that many proteins form complexes. Furthermore, translation appears to organise the chromosome structure, with specific islands corresponding to particular codon usage biases [21]. The consequence is that the amino acid composition of proteins cannot be random, and indeed there is a large bias in amino acid distribution among the different proteins making a proteome. A multivariate analysis of the proteome (correspondence analysis) of a large number of prokaryotes showed that proteins are grouped into clusters comprising a similar distribution in particular amino acids. A strong bias opposes charged residues to hydrophobic residues and permits one to identify with remarkable precision the protein located in the inner membrane of the cell (IIMPs) [30]. Two further biases, apparently universal, characterise the bacterial proteomes. There is a bias, perhaps not unexpected, created by the G+C composition of the genome, and another one, driven by the aromatic composition of the proteins. Interestingly, aromatic-rich proteins are most often without recognised function. This group is also highly enriched in "orphan" proteins [31]. An explanation to this observation is that proteins created *de novo* might indeed go through progressively enhanced functional properties, starting from the general function of stabilising complexes by acting as "gluons", where they use the intrinsic stickiness of aromatic amino acids [30].

A further bias appeared in proteins coded by psychrophilic organisms. Indeed their proteome is systematically enriched in asparagine, while the dioxygensensitive amino acids, cysteine, histidine and methionine are counter-selected [32, 33]. This bias corresponds to intrinsic properties of asparagine, which isomerises easily, leading to perhaps the major post-translational modification in all proteomes. Asparagine spontaneously isomerises in particular contexts into isoaspartate, with concomitant deamidation in a reaction which is still poorly understood. This reaction affects protein structures (and may affect their function). It may also have a role in regulating protein folding and it is a signal for degradation of intracellular proteins [34]. Aspartate and asparagine isomerisation is therefore another physico-chemical constraint that needs to be dealt with using appropriate metabolic patches. In many organisms (including *Escherichia coli* and *Homo sapiens*) there exists a process that can restore aspartate from isoaspartate after methylation and demethylation, an extremely costly repair system [35].

This observation leads us to revisit the inevitability of ageing. Indeed, be it only because of asparagine/aspartate isomerisation, proteins age, sometimes very fast (e. g. ribosomal protein S 11 from *E. coli*, within minutes at 37 °C [36]). As a consequence, it is always an aged cell (or multicellular organism) that gives birth to a young one. This implies that in the process of forming a progeny, there is creation of information. We need therefore to identify the genes acting in the process of accumulating information [10].

REVISITING INFORMATION

A natural way to consider information is to appreciate its "value". This implies intuitively that one will need energy to create information. This was indeed the common view until Rolf Landauer showed in 1961 that creation of information is reversible and therefore does not require any energy [37, 38]. This remarkable work, curiously widely ignored, showed however that reversibility was at a cost: an enormous amount of time or space was required to permit reversible creation of information. Hence, creation of information could only be tolerated if a process existed that permitted to "make room" for novel information to be further created. By contrast with reversibility of the creation of information, this process required consumption of energy.

In this context of physics, improvement of metabolism over time is therefore not an impossibility. It can be at least conceptually tolerated, as creation of information is reversible. However, in order to proceed efficiently, the corresponding process will require a specific process to "make room": how is this obtained? Can we identify in genomes the genes coding for the functions required to put this process into action?

In order to proceed with this investigation, which assumes the existence of a fairly ubiquitous process, we need to look for ubiquitous functions. However, with genome studies we have only direct access to sequences (and sometimes structures), while "acquisitive evolution" systematically masks functional persistence. Briefly, any system submitted to the trio *variation/selection/amplification* will evolve, as it will open windows for novel functions (note that this is creation of information). Functions however can only exist when a concrete object is recruited, so that many objects will fulfil a given function [39, 40]. The consequence is that it is not possible to identify the presumably ubiquitous genes which would correspond to ubiquitous functions, simply because they will not be ubiquitous.

FROM FUNCTIONAL UBIQUITY TO GENE PERSISTENCE

To sum up, functional ubiquity does not imply structural ubiquity. Fortunately, however, living organisms evolve by descent, and efficient objects tend to persist through time because their genes will tend to be conserved over generations. Briefly, there is some kind of stickiness in the adaptation of an object to a particular function. Hence, rather than look for ubiquity, we should look for "persistence", i. e. for the tendency of a gene to be present in a given number of genomes. And looking for persistence will permit us to identify ubiquitous functions. We need to note here that any approach to this quest will heavily depend on the genome sample we possess, making it a fairly difficult enterprise. As in the quest for consensuses in sequences, we expect that the sampling bias will go through a maximum when the number of genomes increases, and then slowly decrease as more genomes are available (the exception makes the rule) [41]. Appropriate computing techniques can be set up to deal with this problem and it is possible to find out persistent genes from the present collection of genome sequences.

With this view, a set of 400 – 500 genes has been identified, that persist in bacterial genomes and, as expected, the vast majority of the genes labelled as "essential" (because they cannot be inactivated without complete loss of viability) belonged to this set [42]. Is, then, "persistent" a synonym of "essential"? A remarkable feature of both categories of genes is that, as in the case of genes identified as essential in the laboratory, most persistent genes are located in the leading replication strand, suggesting that they respond to common selection pressures. In terms of functions the ~250 essential genes code for the bulk of the functions involved in information transfers. The functions in this list are not unexpected, as the list could be established very early on, and was indeed at the root of the interest of the European Commission for sequencing genomes, for example [43]. A list established using the most degenerate autonomous organisms also resulted later in a similar number [44].

The category of non essential persistent genes is interesting, both because it was not predicted by the latter studies, and because it is very much biased in particular functions. It codes for functions involved in stress, maintenance and repair, on the one hand, and for a few metabolic patches, in particular for serine degradation into pyruvate [42], on the other hand. An important feature of this particular set is that it codes for functions that may have a consequence in the long term, and have not, therefore, been studied properly under laboratory conditions. Indeed, studies investigating essentiality have just tested the capacity of mutants to grow and to generate a colony after individual gene inactivation.

CLUSTERING OF PERSISTENT GENES

The location of persistent genes in the leading DNA strand (as does the subcategory of essential genes) combines with another specific feature in their organisation in genomes: they tend to cluster together. Using 228 genomes comprising more than 1,500 genes (to avoid sampling biases) and accurate annotations, we identified genes that tend to remain close to one another. This "mutual attraction" constructed a remarkable network made of three layers, building networks with differing connectivities. These layers can be grouped into a consistent picture in relation to the functional properties of the genes they are made of. A first network, made of genes coding for the construction of the building blocks of intermediary metabolism (nucleotides and coenzymes, lipids), is highly fragmented.

A second network is built around class I tRNA synthetases, and a third network, almost continuous, is organised around genes coding for the ribosome, for transcription and replication and for other functions managing information transfers [5].

How can we account for this clustering? Based on the observation that the gene flow mediated by horizontal gene transfer (HGT) must be high, we proposed that a purely passive clustering process is at work in genomes, noting that what has been interpreted as causes (co-transcription and formation of operons, and protein-protein interactions) are, instead, consequences of clustering. If genes are deleted as local bundles (to compensate for bundles of genes introduced by HGT), this results in a purely passive gene clustering, as the progeny of cells with clustering of the most important genes for sustaining life (persistent genes) are more likely to survive than that of cells with genes uniformly spread in the genome (uniform distribution is the largest deviation from clustering) [45].

A noteworthy feature of this organisation is that it emphasises the separation between metabolism and replication and is consistent with the scenario:

$$\text{Building blocks} \Rightarrow \text{nucleotides} \Rightarrow \text{tRNA} \Rightarrow \text{ribosome} \Rightarrow \text{DNA}$$

which is highly reminiscent of what could have happened at the origin of life. To better understand its meaning, let us briefly explore a mineral scenario for the origin of life [46, 47].

The surface of charged solids (e.g. pyrite (Fe-S) [48]) selects and compartimentalises charged molecules; this first step forms some amino acids, the main coenzymes, fatty acids and ribonucleotides; polymerisation with elimination of water molecules increases entropy and is therefore favoured on surfaces. Subsequently (once the nucleotides have been created) compartmentalised metabolism creates surface substitutes via polymerisation of ribonucleotides in the presence of peptides, with ancestors of tRNA (the RNA world), via a shift of role, from that of a *substrate* to that of a *template*. Then, RNAs develops its template role in translation with the invention of the genetic code, placing the ribosome as the core structure of nascent life. Finally, it further shifts away from its role as a substrate for metabolic reactions, to template for self-replication, discovering the complementarity law. Nucleic acids are further stabilised by the invention of deoxyribonucleotides, at the time when the rules controlling information transfer are discovered, first within the RNA world where vesicles carrying the ancestors of genes split and fuse randomly, before formation of the first genomes.

THE PALEOME AND THE CENOME

Coming back to the organisation of bacterial genomes, we now can see them as composed of two major parts. Persistent genes, with this scenario reminiscent of the origin of life, form the *paleome* (from παλαφος ancient) with genes coding for the basic functions permitting cells to survive and to perpetuate life. Bacteria need also to occupy a particular environment.

In 1877, Karl Möbius referred to the common pool of living species in a particular environment as a *biocenose* (see e.g. [49]). While the concept of gene did not exist at the time, we need now to relate this idea to that of the genes permitting the cell to occupy an ecological niche. These genes are acquired by HGT from a large unknown pool of genes and, as a consequence of the corresponding gene transfer processes (transformation, transduction, integration of prophages and conjugation), they are generally coming in genomes as gene clusters [50]. This very large class, the *cenome* (after κοφνοζ, common, as in biocenose [5]) tends to comprise novel members in different strains of the same species (Figure 3). Taking into account the concept of *pan-genome*, which puts together all the genes of a given species [51], the cenome of a given species is a subset of the pan-genome, comprising all the genes permitting any strain of that species to live in its favoured niche. As stated above, in a species such as *E. coli* the pan-genome is mostly made of genes forming the cenome of each individual strain, and is already larger than 20,000 genes with no sign of levelling off as new strain genomes are sequenced.

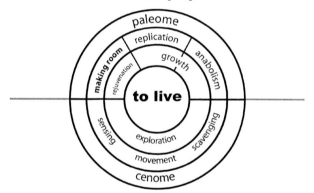

Figure 3. The paleome and the cenome. The paleome codes for functions necessary for survival and for propagation of life. Among its non-essential part one finds genes coding for metabolic patches meant to remedy conditions of metabolic incompatibilities (frustration). Such patches can also be found in the cenome as they are necessary when a pathway produces highly reactive molecules such as alpha-diketones.

PROVISIONAL CONCLUSION: A SPLIT PALEOME

To live associates three major processes: to survive ageing processes, to perpetuate life while already aged, and to live in a particular context. The first two processes require presumably ubiquitous functions, which have been grouped in the set we named the paleome, because of the way it organises relative to a large number of bacterial genomes. Most of the functions in the paleome have been identified, at least at the biochemical level. They are often (as in the case of transcription and translation processes) the target of drugs that prevent propagation of the relevant organisms. While these functions are conserved, they are often not resulting from similar structures, so that they can only be identified via analysis of gene persistence. The functions of the paleome may be split following a variety of specific characters. For example, half of its genes code for functions that are essential to permit formation of a colony on plates supplemented by rich medium: they are the functions of essential genes [52, 53]; the other half, while ubiquitous, does not have this property [42]. This latter half comprises mostly genes that are essential to perpetuate life, but are not essential in the short term [10]. Another split identifies functions which solve some of the metabolic incompatibilities in the cell, resulting from chemical constraints such as spontaneous isomerisation of aspartate and asparagine. This phenomenon of frustration is necessarily quite widespread, as a large number of chemical intermediates, such as alpha-diketones, are extremely reactive towards amino groups. This explains why the downstream section of the methionine salvage pathway, which recycles the methylthioadenosine formed from a variety of reactions derived from S-adenosyl-methionine, is highly variable [54]. The functions in this pathway, typical of what is found in the cenome of an organism, has only been solved in the case of reactions involving dioxygen, while it is certainly present in anaerobic organisms (Figure 4). Its very existence examplifies the type of unknown functions we should look for when exploring the cenome of organisms, in particular in metagenomic studies. This will require considerable imagination for the prediction of novel chemical reactions.

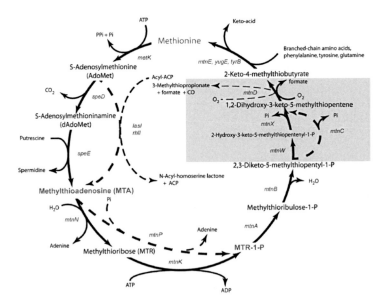

Figure 4. Present knowledge of the methionine salvage pathway. The pathway has been decyphered in [54]. The shaded region corresponds to the reactions downstream from the highly reactive 2,3-Diketo-5- methylthiopentyl-1-phosphate. They are chemically completely different in different organisms (MtnW is a RuBisCO-like enzyme with no relation with the MtnC enolase-phosphatase).

ACKNOWLEDGEMENTS

This work benefited from many years of continuous discussions with the Stanislas Noria group. Support for *in silico* analyses and experiments came from the PROBACTYS programme, grant CT-2006–029104 in an effort to define genes essential for the construction of a synthetic cell and the BioSapiens Network of Excellence, grant LSHG CT-2003–503265.

REFERENCES

[1]   Oliver, S.G., van der Aart, Q.J., Agostoni-Carbone, M.L., Aigle, M., Alberghina, L., Alexandraki, D., Antoine, G., Anwar, R., Ballesta, J.P., Benit, P., *et al.* (1992) The complete DNA sequence of yeast chromosome III. *Nature* **357**:38–46.

[2]   Glaser, P., Kunst, F., Arnaud, M., Coudart, M.P., Gonzales, W., Hullo, M.F., Ionescu, M., Lubochinsky, B., Marcelino, L., Moszer, I., Presecan, E., Santana, M., Schneider, E., Schweizer, J., Vertes, A., Rapoport, G., Danchin, A. (1993) *Bacillus subtilis* genome project: cloning and sequencing of the 97 kb region from 325 degrees to 333 degrees. *Mol. Microbiol.* **10**:371–384.

[3]   Danchin, A. (2003). *The Delphic boat. What genomes tell us.* Trans. A. Quayle. Harvard University Press, Cambridge (Mass, USA).

[4]   de Lorenzo, V., Danchin, A. (2008) Synthetic biology: discovering new worlds and new words. The new and not so new aspects of this emerging research fields. *EMBO Reports* **9**:822–827.

[5]   Danchin, A., Fang, G., Noria, S. (2007) The extant core bacterial proteome is an archive of the origin of life. *Proteomics* **7**:875–889.

[6]   Hofstadter, D. (1979). "Gödel, Escher, Bach: an Eternal Golden Braid". Basic Books, New York.

[7]   Muller, H. (1932) Some genetic aspects of sex. *The American Naturalist* **66**:118–128.

[8]   Orgel, L. (1963) The maintenance of the accuracy of protein synthesis and its relevance to aging. *Proc. Natl. Acad. Sci. U.S.A.* **49**:517–521.

[9]   Dyson, F. J. (1985). *Origins of life.* Cambridge University Press, Cambridge, UK.

[10]   Danchin, A. (2008) Natural Selection and Immortality. *Biogerontology (submitted).*

[11]   Cover, T., Thomas, J. (1991). *Elements of information theory.* Wiley, New York.

[12]   Bennett, C. (1988) Logical Depth and Physical Complexity. In *The Universal Turing Machine: a Half-Century Survey* (R. Herken, ed.), pp. 227–257. Oxford University Press, Oxford.

[13]   Danchin, A. (1996) On genomes and cosmologies. In *Integrative Approaches to Molecular Biology* (J. Collado-Vides, B. Magasanik, T. Smith, eds.), pp. 91–111. The MIT Press, Cambridge (USA).

[14]   Turing, A. (1936–1937) On computable numbers, with an application to the Entscheidungsproblem. *Proceedings of the London Mathematical Society* **42**:230–265.

[15]   Turing, A. (1946 (1986)) A. M. Turing's ACE Report of 1946 and Other Papers. *In* Charles Babbage Institute reprint series for the History of Computing (B. Carpenter, R. Doran, eds.), Vol. 10. MIT Press, Cambridge (Mass).

[16]   von Neumann, J. (1958 (reprinted 1979)). *The Computer and the Brain*. Yale University Press, New Haven.

[17]   Médigue, C., Rouxel, T., Vigier, P., Hénaut, A., Danchin, A. (1991) Evidence for horizontal gene transfer in *Escherichia coli* speciation. *J. Mol. Biol.* **222**:851–856.

[18]   Lartigue, C., Glass, J.I., Alperovich, N., Pieper, R., Parmar, P.P., Hutchison, C.A., 3rd, Smith, H.O., Venter, J.C. (2007) Genome transplantation in bacteria: changing one species to another. *Science* **317**:632–638.

[19]   Bussiere, D.E., Bastia, D. (1999) Termination of DNA replication of bacterial and plasmid chromosomes. *Mol. Microbiol.* **31**:1611–1618.

[20]   Thanaraj, T.A., Argos, P. (1996) Ribosome-mediated translational pause and protein domain organization. *Protein Sci.* **5**:1594–1612.

[21]   Bailly-Bechet, M., Danchin, A., Iqbal, M., Marsili, M., Vergassola, M. (2006) Codon usage domains over bacterial chromosomes. *PLoS Comput Biol* **2**:e37.

[22]   Averof, M., Akam, M. (1995) Hox genes and the diversification of insect and crustacean body plans. *Nature* **376**:420–423.

[23]   Averof, M. (1997) Arthropod evolution: same Hox genes, different body plans. *Curr. Biol.* **7**:R634–636.

[24]   Tamames, J., Gonzalez-Moreno, M., Mingorance, J., Valencia, A., Vicente, M. (2001) Bringing gene order into bacterial shape. *Trends Genet.* **17**:124–126.

[25]   Kitao, A., Yonekura, K., Maki-Yonekura, S., Samatey, F.A., Imada, K., Namba, K., Go, N. (2006) Switch interactions control energy frustration and multiple flagellar filament structures. *Proc. Natl. Acad. Sci. U.S.A.* **103**:4894–4899.

[26]   Gibson, D.G., Benders, G.A., Andrews-Pfannkoch, C., Denisova, E.A., Baden-Till-son, H., Zaveri, J., Stockwell, T.B., Brownley, A., Thomas, D.W., Algire, M. A., Merryman, C., Young, L., Noskov, V.N., Glass, J.I., Venter, J.C., Hutchison, C.A., 3 rd, Smith, H.O. (2008) Complete chemical synthesis, assembly, and cloning of a *Mycoplasma genitalium* genome. *Science* **319**:1215 – 1220.

[27]   Rocha, E., Danchin, A., Viari, A. (1999) Universal replication biases in bacteria. *Mol. Microbiol.* **32**:11 – 16.

[28]   Rocha, E., Danchin, A. (2003) Gene essentiality determines chromosome organisa-tion in bacteria. *Nucleic Acids Res.* **31**:6570 – 6577.

[29]   Rocha, E.P., Danchin, A. (2004) An analysis of determinants of amino acids sub-stitution rates in bacterial proteins. *Mol. Biol. Evol.* **21**:108 – 116.

[30]   Pascal, G., Médigue, C., Danchin, A. (2005) Universal biases in protein composition of model prokaryotes. *Proteins* **60**:27 – 35.

[31]   Pascal, G., Médigue, C., Danchin, A. (2006) Persistent biases in the amino acid composition of prokaryotic proteins. *Bioessays* **28**:726 – 738.

[32]   Riley, M., Staley, J.T., Danchin, A., Wang, T.Z., Brettin, T.S., Hauser, L.J., Land, M.L., Thompson, L.S. (2008) Genomics of an extreme psychrophile, *Psychromonas ingrahamii*. *BMC Genomics* **9**:210.

[33]   Médigue, C., Krin, E., Pascal, G., Barbe, V., Bernsel, A., Bertin, P.N., Cheung, F., Cruveiller, S., D'Amico, S., Duilio, A., Fang, G., Feller, G., Ho, C., Mangenot, S., Marino, G., Nilsson, J., Parrilli, E., Rocha, E.P., Rouy, Z., Sekowska, A., Tutino, M.L., Vallenet, D., von Heijne, G., Danchin, A. (2005) Coping with cold: the genome of the versatile marine Antarctica bacterium *Pseudoalteromonas halo-planktis* TAC 125. *Genome Res.* **15**: 1325 – 1335.

[34]   Shimizu, T., Matsuoka, Y., Shirasawa, T. (2005) Biological significance of isoaspar-tate and its repair system. *Biol. Pharm. Bull.* **28**:1590 – 1596.

[35]   Clarke, S. (2003) Aging as war between chemical and biochemical processes: protein methylation and the recognition of age-damaged proteins for repair. *Ageing Res. Rev.* **2**:263 – 285.

[36]   David, C.L., Keener, J., Aswad, D.W. (1999) Isoaspartate in ribosomal protein S 11 of *Escherichia coli*. *J. Bacteriol.* **181**:2872 – 2877.

[37]   Landauer, R. (1961) Irreversibility and heat generation in the computing process. *IBM Journal of research and development* **3**:184 – 191.

[38]   Bennett, C. (1988) Notes on the history of reversible computation. *IBM Journal of research and development* **44**:270 – 277.

[39]   Thompson, L.W., Krawiec, S. (1983) Acquisitive evolution of ribitol dehydrogenase in *Klebsiella pneumoniae*. *J. Bacteriol.* **154**:1027–1031.

[40]   Ashida, H., Danchin, A., Yokota, A. (2005) Was photosynthetic RuBisCO recruited by acquisitive evolution from RuBisCO-like proteins involved in sulphur metabolism? *Res. Microbiol.* **156**:611–618.

[41]   Hénaut, A., Danchin, A. (1996) Analysis and predictions from *Escherichia coli* sequences or *E. coli in silico*. In *Escherichia coli and Salmonella, Cellular and Molecular Biology* (F. Neidhardt, ed.), Vol. 1, pp. 2047–2065. ASM Press, Washington.

[42]   Fang, G., Rocha, E., Danchin, A. (2005) How essential are nonessential genes? *Mol. Biol. Evol.* **22**:2147–2156.

[43]   Danchin, A. (1988) Complete genome sequencing: future and prospects. In *BAP 1988–1989* (A. Goffeau, ed.), pp. 1–24. Commission of the European Communities, Brussels.

[44]   Mushegian, A.R., Koonin, E.V. (1996) A minimal gene set for cellular life derived by comparison of complete bacterial genomes. *Proc. Natl. Acad. Sci. U.S.A.* **93**:10268–10273.

[45]   Fang, G., Rocha, E.P., Danchin, A. (2008) Persistence drives gene clustering in bacterial genomes. *BMC Genomics* **9**:4.

[46]   Granick, S. (1957) Speculations on the origins and evolution of photosynthesis. *Ann. N. Y. Acad. Sci.* **69**:292–308.

[47]   Danchin, A. (1989) Homeotopic transformation and the origin of translation. *Prog. Biophys. Mol. Biol.* **54**:81–86.

[48]   Wächtershäuser, G. (1988) Before enzymes and templates: theory of surface metabolism. *Microbiol. Rev.* **52**:452–484.

[49]   Movila, A., Uspenskaia, I., Toderas, I., Melnic, V., Conovalov, J. (2006) Prevalence of *Borrelia burgdorferi* sensu lato and *Coxiella burnetti* in ticks collected in different biocenoses in the Republic of Moldova. *International Journal of Medical Microbiology* **296**:172–176.

[50]   Lawrence, J.G., Roth, J.R. (1996) Selfish operons: horizontal transfer may drive the evolution of gene clusters. *Genetics* **143**:1843–1860.

[51]   Tettelin, H., Masignani, V., Cieslewicz, M.J., Donati, C., Medini, D., Ward, N.L., Angiuoli, S.V., Crabtree, J., Jones, A.L., Durkin, A.S., Deboy, R.T., Davidsen, T.M., Mora, M., Scarselli, M., Margarit y Ros, I., Peterson, J.D., Hauser, C.R., Sundaram, J.P., Nelson, W.C., Madupu, R., Brinkac, L.M., Dodson, R.J., Rosovitz, M.J., Sulli-

van, S.A., Daugherty, S.C., Haft, D.H., Selengut, J., Gwinn, M.L., Zhou, L., Zafar, N., Khouri, H., Radune, D., Dimitrov, G., Watkins, K., O'Connor, K.J., Smith, S., Utterback, T.R., White, O., Rubens, C.E., Grandi, G., Madoff, L.C., Kasper, D.L., Telford, J.L., Wessels, M.R., Rappuoli, R., Fraser, C.M. (2005) Genome analysis of multiple pathogenic isolates of *Streptococcus agalactiae*: implications for the microbial "pangenome". *Proc. Natl. Acad. Sci. U.S.A.* **102**:13950–13955.

[52] Kobayashi, K., Ehrlich, S.D., Albertini, A., Amati, G., Andersen, K.K., Arnaud, M., Asai, K., Ashikaga, S., Aymerich, S., Bessieres, P., Boland, F., Brignell, S.C., Bron, S., Bunai, K., Chapuis, J., Christiansen, L.C., Danchin, A., Debarbouille, M., Dervyn, E., Deuerling, E., Devine, K., Devine, S.K., Dreesen, O., Errington, J., Fillinger, S., Foster, S.J., Fujita, Y., Galizzi, A., Gardan, R., Eschevins, C., Fukushima, T., Haga, K., Harwood, C.R., Hecker, M., Hosoya, D., Hullo, M.F., Kakeshita, H., Karamata, D., Kasahara, Y., Kawamura, F., Koga, K., Koski, P., Kuwana, R., Imamura, D., Ishimaru, M., Ishikawa, S., Ishio, I., Le Coq, D., Masson, A., Mauel, C., Meima, R., Mellado, R.P., Moir, A., Moriya, S., Nagakawa, E., Nanamiya, H., Nakai, S., Nygaard, P., Ogura, M., Ohanan, T., O'Reilly, M., O'Rourke, M., Pragai, Z., Pooley, H.M., Rapoport, G., Rawlins, J.P., Rivas, L.A., Rivolta, C., Sadaie, A., Sadaie, Y., Sarvas, M., Sato, T., Saxild, H.H., Scanlan, E., Schumann, W., Seegers, J.F., Sekiguchi, J., Sekowska, A., Seror, S.J., Simon, M., Stragier, P., Studer, R., Takamatsu, H., Tanaka, T., Takeuchi, M., Thomaides, H.B., Vagner, V., van Dijl, J.M., Watabe, K., Wipat, A., Yamamoto, H., Yamamoto, M., Yamamoto, Y., Yamane, K., Yata, K., Yoshida, K., Yoshikawa, H., Zuber, U., Ogasawara, N. (2003) Essential *Bacillus subtilis* genes. *Proc. Natl. Acad. Sci. U.S.A.* **100**:4678–4683.

[53] Joyce, A.R., Reed, J.L., White, A., Edwards, R., Osterman, A., Baba, T., Mori, H., Lesely, S.A., Palsson, B.O., Agarwalla, S. (2006) Experimental and computational assessment of conditionally essential genes in *Escherichia coli*. *J. Bacteriol.* **188**:8259–8271.

[54] Sekowska, A., Denervaud, V., Ashida, H., Michoud, K., Haas, D., Yokota, A., Danchin, A. (2004) Bacterial variations on the methionine salvage pathway. *BMC Microbiol* **4**:9.

CATALYSIS AT THE ORIGIN OF LIFE VIEWED IN THE LIGHT OF THE (*M,R*)-SYSTEMS OF ROBERT ROSEN

ATHEL CORNISH-BOWDEN* AND MARÍA LUZ CÁRDENAS

Unité de Bioénergétique et Ingénierie des Protéines, Centre National de la Recherche Scientifique, 31 chemin Joseph-Aiguier, B.P. 71, 13402 Marseilles, France

E-Mail: *acornish@ibsm.cnrsmrs.fr

Received: 1st September 2008 / Published: 16th March 2009

ABSTRACT

Living systems as we know them today are both complex, displaying emergent properties, and extremely complicated, with huge numbers of different components. At the origin of life they must also have had emergent properties, and hence must have been complex, but they cannot have been as complicated as modern organisms, because we cannot imagine that the first organisms started with anything as elaborate as a ribosome and all of the protein-synthesis machinery. Understanding how complexity could arise in even the simplest early organism requires, however, a theory of life, something that is largely lacking from modern biology. Various authors have contributed elements of such a theory, and the *(M,R)*-systems of Robert Rosen provide a convenient starting point.

INTRODUCTION

In an earlier contribution to this series [1] we commented that many phenomena are described as complex when in reality they are no more than complicated, because they can be fully accounted for in terms of the properties of their components: there is no "emergence". It must be recognized, however, that it is not always easy to decide whether a property is truly emergent or not, in part because of disagreements about how emergence should be defined [2]. Living organisms in their totality, however, are complex, because it appears to be impossible to deduce their properties solely by applying the reductionist programme of studying all of the properties of all of the components in sufficient detail. Some authors [3, 4] go further, and say that it not only appears to be impossible but it really is impossible even in principle, but this aspect remains controversial [5] and we shall not discuss it here.

We shall, however, try to resume the current state of understanding of the nature of life and the definition of a living organism. Although this might seem an essential component of biology, it is in practice ignored by nearly all biologists [6] and regarded as irrelevant to the practice of modern biology by many [7], as we discussed previously [8]. The famous question raised by Erwin Schrödinger [9] of what life is remains unanswered more than 60 years later. Parts of it, of course, have been answered: we can identify Schrödinger's "codescript" with the DNA in which protein sequences are recorded; we now rarely need to speak of organisms feeding on negative entropy because we understand that living organisms are thermodynamically open systems that can maintain themselves far from equilibrium without violating any thermodynamic principles. Nonetheless, the crucial question of what biological organization actually means and how it is maintained almost indefinitely remains inadequately studied. In 2005 the editors of *Science* [10] celebrated 125 years of existence of the magazine with a list of 125 questions, "the most compelling puzzles and questions facing scientists today". A high proportion of these were questions about biology, and included such vogue items as "is an effective HIV vaccine feasble?", but Schrödinger's question was not among them.

The first modern attempts to understand biological organization were made by Stéphane Leduc [11]. His osmotic experiments produced impressively complicated and biological-looking structures (which can be reproduced in full colour illustrations today: see Querbes [12]), but few biologists today would accept that osmosis tells us much about the forms of real organisms. However, his more general belief that natural selection is not the only explanation of biological forms, and that chemical reactivity as well as physicochemical and mechanical forces also play major roles remains important, and was taken up by D'Arcy Thompson [13] in a closely argued book that has been very influential in modern thinking.

There are four principal current theories that try to explain biological organization, the (M,R)-systems of Rosen [3], the chemoton of Gánti [14], the autopoiesis of Maturana and Varela [15], and the autocatalytic sets of Kauffman [16]. Despite the fact that all of these theories contain some of the same ideas, they are by no means the same as one another, and none of the authors mentioned makes any reference to any of the others in their principal publications. Although all attach importance to "closure" and their definitions of this overlap, they underline different aspects. Rosen [3], for example, refers to closure to efficient causation, which means that all the catalysts required by an organism need to be products of its own metabolism; Maturana and Varela [15] stress structural closure, or the need for an organism to be enclosed within a membrane, cell wall, or skin; Kauffman [16] considers that catalytic closure is the consequence of very large sets of different polypeptides or polynucleotides; Gánti [14] agrees on the necessity for structural closure, and also emphasizes the need for any theory of life to be rooted in an adequate knowledge of chemistry. Clearly, therefore, an important task for the future will be to integrate all of these threads into a single theory of life. Here we shall be less ambitious, concentrating on the ideas of Robert Rosen, which are the most abstract and difficult to understand of those we have mentioned, and will use them to analyse aspects of catalysis at the origin of life.

LIMITS OF REDUCTIONISM

The reductionist approach has taken biochemistry a very long way since Buchner [17] first demonstrated that alcoholic fermentation could occur in a cell-free extract of yeast, and it is very unlikely that we should know much about biochemistry today, and still less about molecular biology, if the approach in the 20th century had not been overwhelmingly reductionist. However, it is one thing to recognize the progress that reductionism has brought, but it is another to suppose that this can continue indefinitely. One can certainly understand the behaviour of components of the cell, such as metabolic pathways, in terms of their components, enzymes in this case, and enzymes can be understood in terms of the properties of their side-chains. However, especially at the level of the cell or the whole organism, the reductionist approach cannot provide the whole truth, because these entities have complex and emergent properties. However, today, and in contrast to the early 20th century, we do try to understand the chemistry of whole organisms (systems chemistry) and, in the words of Henrik Kacser [18], "one thing is certain: to understand the whole you must study the whole", an idea more picturesquely expressed by a Russian proverb, "a hundred rabbits do not make a horse".

It may be illuminating to compare modern biology with modern physics. In the 19th century theory followed from experimental observations: thermodynamics, for example, developed from Sadi Carnot's efforts to determine whether steam engines could be improved without limit. In this and other 19th century cases the experiments preceded the development of the theory, but in 20th century physics theory usually preceded experiment: Albert Einstein, for example, did not develop the theory of relativity after wondering how the satellite navigation system in his car worked; on the contrary, this and other applications of relativity came many years after the theory had been worked out. The comparison is not just with physics, and as Günter von Kiedrowski remarked earlier in this symposium, "a century ago chemistry was in the same situation as biology today". We believe, in summary, that future advances in biology will require a more complete theoretical basis than is provided by the theory of natural selection, the only general biological theory that exists, which is valuable for interpreting observations, but is not the whole truth.

CLOSURE TO EFFICIENT CAUSATION

Metabolism is often represented as a large and complicated set of processes catalysed by enzymes or transporters, these processes including both chemical reactions and transport across membranes. This description, however, while true as far as it goes, is seriously incomplete. As biologists know, and as Rosen [3] emphasized, the enzymes and transporters (which for brevity we shall consider together just as enzymes) are subject to turnover, dilution by growth of the organism, and losses due to their finite stability, processes that we shall abbreviate to decay. So, even if all the necessary enzymes are present in one moment, they cannot continue to be present indefinitely unless mechanisms exist to replace them. From what, however, can they be made? Clearly the only possible answer is that they must be made from the products of metabolism, so they are themselves products of metabolism, and hence metabolites.

Not only must all "enzymes" be considered metabolites in this sense, but many "metabolites" are also enzymes, because they are biological molecules that act as catalysts: metabolic cycles require not only the protein catalysts usually regarded as enzymes, but also the molecules that are consumed and regenerated in the process: the urea cycle, for example, requires not only three proteins, citrullinase, arginine deaminase and arginase, but also ornithine, and so ornithine has just as much right to be called an enzyme as the three proteins. It has more right, even, as the cycle would still occur (slowly) if some or all of the proteins were missing, but it would not occur without ornithine or a molecule that could replace it, such as citrulline or arginine. It follows that the usual distinction between enzymes and metabolites is formally meaningless [19]. Similar considerations apply to other metabolic cycles, such as the tricarboxylate cycle.

To refer to this organized replacement Rosen [3] used the unfortunate term repair, inviting confusion with more standard notions of repair in modern biochemistry, such as DNA repair and action of chaperones, so we prefer to refer to *replacement* [20, 21]. Fortunately this begins with the same letter of the alphabet as Rosen's word, so we can continue to use the term (M,R)-system as a short form of *metabolism-replacement system*, which summarizes Rosen's view of an organism. The essential point is that catalysts need to be replaced internally, by the organism itself; they cannot be replaced by an external agency. For this reason an organism is fundamentally different from a machine, because regardless of how one defines a machine, whether a simple tool such as an axe or something as complicated as an airliner, or even as a complete factory, at no level of definition does the machine make itself or maintain itself. The machine analogy may be helpful for understanding certain properties of organism, for example how the heart work, but in general it fails, because an organism is not a machine.

Unfortunately, however, Rosen did not make it easy for his readers to study his work. He presented it in resolutely mathematical terms, making no concessions to readers without mathematical expertise, he provided no examples to illustrate his central points – not even mathematical examples, and certainly no biological examples, and he did not define the range of validity of his ideas. We have therefore tried to fill these voids [20, 21].

CATALYSIS AT THE ORIGIN OF LIFE

At the origin of life there was no natural selection as we understand it today, but there was certainly chemical selection resulting from differences in rates of reactions derived from kinetic or thermodynamic properties [22]. Thus (*M,R*)-systems (*metabolism-replacement-systems*) probably emerged in prebiotic conditions thanks to the presence of inorganic catalysts or simple organic molecules that could act as catalysts. Modern organisms are not only complex; they are also extremely complicated, with a wide array of regulatory mechanisms, both metabolic and genetic, that were surely absent at the origin of life. These mechanisms are not explicitly visible in (*M,R*)-systems (though they are not excluded, and can be considered to be implicit), and so the representation of an organism as an (*M,R*)-system may be closer to the reality at the origin of life than to the reality of today.

Catalysis is fundamental for the organization of living systems, and must have been necessary at the origin of life, to permit organized systems to appear, to maintain themselves, and to grow. Some degree of specificity was also necessary, to allow one system to be different from another. Thus although the first catalysts must have been much simpler molecules than the protein or RNA catalysts that we know today, they must have had properties closer to these than to highly unspecific catalysts like platinum black. Specificity could then have developed progressively, first through chemical evolution and then through natural selection, to arrive finally at present-day bio-catalysts. However, specificity cannot be complete, because it is not possible for a system to fabricate its own catalysts if each one needs its own unique catalyst.

TYPES OF CLOSURE

The idea of closure may be understood in three fundamentally different ways, as illustrated in Figure 1. The simplest is the structural closure (Fig. 1a) produced by the physical boundary (membranes, cell walls, skin) that encloses an individual: in a present-day organism this is always fabricated by the organism itself, but this may not have been true for the first organisms, which could perhaps have made use of already existing inorganic compartments. Organisms must be closed in this sense, because every individual must be distinguishable from every other, and it is in general clear where one individual ends and another begins (if we exclude consideration of *Dictyostelium discoideum* and other organisms that challenge any attempt at a simple definition of an individual). Note that there can be no competition between individuals, and no way of assigning an identity to an individual, without structural closure. (This does not of course exclude the possibility that one individual may live inside another: the more than 10^{14} bacterial residents of a human body are all clearly distinct from one another and from their hosts.) Structural closure forms an essential characteristic of autopoiesis [15] and the chemoton [14], but it was given little or no emphasis by Rosen, who was more concerned with chemical and organizational considerations, which he called *material* and *efficient causation* (terms derived from Aristotle's analysis of causation). We return to the question of individual identity in the section on Individual Identity.

Organisms are *open to material causation*, which simply means that they use chemical molecules taken from their environment that are different from those they excrete into the environment (Fig. 1b). They use the conversion of food (and light, in the case of photosynthetic organisms) into higher-entropy excreta to maintain themselves in states far from equilibrium, and this is a thermodynamic necessity that has been well understood since Schrödinger [9] introduced the idea that organisms "feed on negative entropy".

This thermodynamic necessity does not conflict in any way with Rosen's view that organisms are *closed to efficient causation*, which simply means that they make their own catalysts (Fig. 1c), because the two levels of causation are independent of one another: a system may be closed to one and open to the other without any contradiction.

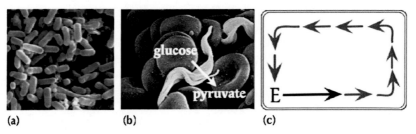

(a)   **(b)**   **(c)**

Figure 1. Three kinds of closure. (a) *Structural closure*. Any individual organism (here illustrated by a culture of *Escherichia coli*) is structurally closed, in the sense that it is separated from all other individuals by a physical barrier, such as a cell wall. (b) *No material closure*. No organism is a closed system in the thermodynamic sense. This is illustrated here by the parasite *Trypanosoma brucei* in the presence of erythrocytes, which can, to a first approximation, be regarded as a small chemical factory that transforms glucose into pyruvate. (c) *Organizational closure*. All organisms are closed in the sense that the catalysts that they need are products of their own metabolism.

Rosen's Relational Diagram

Rosen [3] summarized his view of an organism with a diagram topologically equivalent to that shown in Figure 2. Although at first sight this diagram is unintelligible, it can be understood by following the various steps in a clockwise direction (as we explained earlier [8] in the context of Rosen's own unsymmetrical version of Figure 2), starting with A → B, which represents the whole of metabolism, as full arrows represent material causation: the substrates A of the whole set of metabolic reactions are the material cause of the products B of the same set of reactions. The broken arrows represent efficient causation, and so $f \dashrightarrow$ A $\rightarrow$ B means that metabolism A $\rightarrow$ B is catalysed by the whole set of catalysts f. These catalysts are replaced from the only pool available, the set of metabolic products B under the influence of a replacement system Φ, so $\Phi \dashrightarrow$ B $\rightarrow f$.

This system must itself be replaced, because it is subject to the same problems of decay as the metabolic catalysts. At this point Rosen avoided the incipient infinite regress by supposing that the efficient cause of this replacement of Φ from f could be caused by an entity β that was considered as a property of B, so B $\dashrightarrow f \rightarrow \Phi$ allows the whole diagram to be closed to efficient causation, though, as noted above, it is open to material causation, with a net non-cyclic (or only partially cyclic) process A $\rightarrow$ B. Rosen always made it clear in his writing, and especially in his papers of 1966 and 1971 [24, 25], that although β is related to B it is not the *same* as B, and in mathematical terms it is best understood as the inverse of B [20], but he was less careful to be clear about this in the diagram on which Figure 2 is based, an oversight that has caused a great deal of confusion and misunderstanding in the literature.

The whole diagram is thus closed to efficient causation, or in other words it shows metabolic circularity or organizational invariance. It contains no *final causes* – no explanations of *purpose*, or of *why* anything happens as it does. The fourth of Aristotle's categories of causation, the *formal cause* (what makes a metabolite a metabolite? What makes an enzyme an enzyme?), is also absent from the diagram, as it played little role in Rosen's thinking.

We have already noted that Rosen's term "repair" has nothing to do with conventional uses of this term in modern biochemistry. Perhaps even more misleading, his term "replication" for what we call organizational invariance has *nothing* to do with DNA replication, etc. We include both of these terms in Figure 2 to facilitate comparison with Rosen's publications, but they are otherwise best avoided.

Recent years have seen an enormous growth in "systems biology", with greatly increased interested in "small-world" models of metabolism [26, 27] and the "bow-tie model" of metabolic regulation [28]. However, the idea of metabolic closure is completely absent from these. The small-models are completely concerned with material causation and do not address the question of where the catalysts come from, and although the bow-tie model does imply the existence of catalysts it also does not address the question of where they come from. So although Figure 1 of Csete & Doyle [28], for example, contains numerous feedback loops it has a clear left-to-right reading direction, with no suggestion of closure.

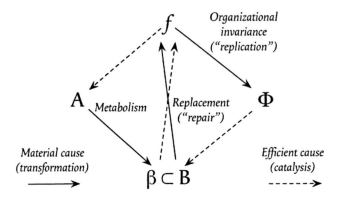

Figure 2. Closure to efficient cause (metabolic circularity). The diagram is based on Figure 10C.6 of Rosen [3], redrawn in the more symmetrical way suggested by Cottam *et al.* [23]. Broken arrows show efficient causes, or catalysis, whereas full arrows represent material causes, or chemical transformations. The diagram is highly abstract: the transformation $f \dashrightarrow A \rightarrow B$ represents the whole of metabolism; the transformation $\Phi \dashrightarrow B \rightarrow f$ represents replacement of enzymes from available products of metabolism, and $\beta \subset B \dashrightarrow f \rightarrow \Phi$ represents the processes need to maintain organizational invariance. Rosen's terms "repair" and "replication" shown in parentheses for these last two processes are misleading, as they have *nothing* to do with the ordinary uses of these words in modern biochemistry, for example for DNA repair and DNA replication. Note that although β is related to B it is not the same as B.

REPRESENTING ROSEN'S MODEL IN A MORE BIOLOGICAL WAY

As noted, Rosen's representation as illustrated in Figure 2 is highly abstract, and not easy to relate to ordinary ideas of biological models. In an attempt to make it more concrete, therefore, we extended a model outlined by Morán *et al.* [29] to arrive at the minimal biological model of an (M,R)-system illustrated in Figure 3 [8, 19–21]. In this representation the three catalytic processes are shown as three cycles of chemical reactions:

1. $S + STU \rightarrow STUS; STUS + T \rightarrow STUST \rightarrow STU + ST$ (metabolism)

2. $ST + SU \rightarrow SUST; SUST + U \rightarrow SUSTU \rightarrow SU + STU$ (replacement)

3. $S + STU \rightarrow STUS; STUS + U \rightarrow STUSU \rightarrow STU + SU$ (organizational invariance)

These can alternatively be written as three catalysed reactions:

$$S + T \xrightarrow{STU} ST, ST + U \xrightarrow{SU} STU, S + U \xrightarrow{STU} SU$$

but the greater simplicity is only apparent, because writing the cycles as catalysed reactions just hides the chemical reality of what is happening. An important point to note here is that closure was achieved by requiring one molecule STU to catalyse two different processes.

We believe that this will be generally true, that for models of arbitrary complexity it will always be necessary to include multifunctional catalysts if closure is to be possible. This in turn implies a general principle: that "moonlighting", or multifunctionality of proteins, as well exemplified by the many different functions of glyceraldehyde 3-phosphate dehydrogenase, is not simply an interesting property of biochemical systems that is being increasingly observed [30, 31], but is an absolute necessity for life. In present-day organisms the ribosome, which participates in the synthesis of a wide variety of different types of protein, certainly contributes to the closure though it is far from being the only example. For consideration of the organisms shortly after the origin of life we need to envisage vastly simpler solutions than that represented by the ribosome and modern catalytic proteins.

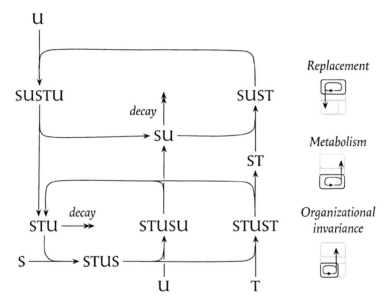

Figure 3. A model of an (*M,R*)-system. This is the biological model suggested previously [8, 19–21]. The two decay reactions shown with double arrowheads are considered to be irreversible. Unfortunately, Figure 3b of [21], which should have been very similar to this one, was printed incorrectly, as three intermediates were named incorrectly, STUST as SUST, STUSU as SUSU, and SUSTU also as SUSU (with the result that the names SUST and SUSU occurred twice each). It was printed correctly in the other papers that have similar diagrams [8, 19].

Another important point to note is that the "solution" to the problem of closure shown in Figure 3 is not unique, because the assignment of STU, SU and again STU as catalysts of the three processes is just one of the $3^3 = 27$ ways in which the three possible catalysts STU, ST and SU could be distributed among the three reactions. Thus the structure of the network does not by itself define Rosen's function β needed for organizational invariance.

A possible way out of the difficulty comes from reexamining the definition of an enzyme or of a catalyst. Any catalysed reaction can be written as a cycle of uncatalysed reactions, as was done in the preceding section, and, as noted in the section Closure to Efficient Causation, it is only convention, not logical necessity, that leads us to called arginase but not ornithine a catalyst of the urea cycle, for example. Catalysis is nothing more fundamental than a human interpretation of chemical cycles, and if this is recognized then Figure 3 represents not a set of three catalysed reactions, but a set of eight uncatalysed chemical reactions, and on this interpretation the question of which catalyst catalyses which reaction does not arise [19]. This means that the catalytic properties follow simply from chemical reactivities, and makes it easier to understand how an organism can "know" which molecule, whether catalyst or other metabolite, is needed for maintaining its metabolism.

INDIVIDUAL IDENTITY

Natural selection implies competition between individuals, and needs identifiable and distinguishable individuals to do the competing. However, Figure 3 does not provide any indication of how the system is contained, and thus "individual", and does not suggest how one individual might be distinguished from another. An obvious step to overcome the first objection will be to incorporate the obligatory formation of a membrane, already explicitly included in autopoiesis [15] and the chemoton [14], into the representation of (M,R)-systems. We have discussed elsewhere [32] how the second objection may be overcome. Although this was in terms of a model somewhat more complicated that the one in Figure 3, the main idea can still be presented in terms of Figure 3. First we need to suppose that the whole system is contained within a barrier (structural closure), as without this there can be no competition. Suppose now that a chance accident causes one of the catalysts, for example SU, to appear in a variant form SU′ that has similar properties to the normal form but catalyses the production of a slightly different form of another catalyst (which must be STU in this case as that is the only possibility). Although initially there may be only one molecule of SU′ the number will increase if it happens that the variant system reacts faster than the normal one and will eventually replace it.

CONCLUSION

Future development of systems chemistry and systems biology will need what Woese [33] has called a guiding vision: the main theories about the nature of life [3, 14–16], in particular the (M,R)-systems of Robert Rosen could constitute this guiding vision. For example, current research efforts to define the minimal genome that allows a system to be autonomous should benefit greatly from the incorporation of such concepts as metabolic closure and organizational invariance.

Efforts must be made to integrate the theories about the nature of life in a single coherent one, which could serve as a basis for understanding how life originated and how it is maintained. Catalysis at the origin of life probably made some use of external agents, so prebiotic systems probably did not fully satisfy the criterion of metabolic closure, but only when they acquired it did they become truly alive, even if the structural closure still depended on external inorganic support. It is tempting therefore to speculate that metabolic closure was acquired before structural closure.

REFERENCES

[1]   Cornish-Bowden, A., Cárdenas, M.L. (2003) Metabolic analysis in drug design: complex, or just complicated? In: *Molecular Informatics: Confronting Complexity* (ed. M. Hicks & C. Kettner), pp. 95 – 107, Beilstein-Institut, Frankfurt.

[2]   Bedau, M.A. (2003) Downward causation and autonomy in weak emergence. *Principia* **6**:5 – 50.

[3]   Rosen, R. (1991) *Life itself: A comprehensive inquiry into the nature, origin and fabrication of life*. New York, Columbia University Press.

[4]   Louie, A.H. (2007) A living system must have noncomputable models. *Artif. Life* **13**:293 – 297.

[5]   Stewart, J., Mossio, M. (2007) Is "life" computable? Proc. ENACTIVE/07 (Proceedings of the 4th International Conference on Enactive Interfaces 2007, Grenoble, France), http://www.enactive2007.org.

[6]   Harold, F. M. (2001) *The Way of the Cell*, Oxford University Press.

[7]   Atlan, H., Bousquet, C. (1994) *Questions de Vie*, Le Seuil, Paris.

[8]   Cornish-Bowden, A., Cárdenas, M.L. (2007) Bring chemistry to life: what does it mean to be alive? In: *Molecular Interactions: Bringing Chemistry to Life* (ed. M. Hicks & C. Kettner), pp. 27 – 39, Beilstein-Institut, Frankfurt.

[9]   Schrödinger, E. (1944) *What is Life?* Cambridge University Press, Cambridge.

[10]   *Science* **309**: 1 – 204 (2005).

[11]   Leduc, S. (1912) *La Biologie Synthétique*, Poinat, Paris.

[12]   Querbes, S. (2007) http://www.stephanequerbes.com/books/recherche/recherche.html.

[13]   Thompson, D'A.W. (1961) *On Growth and Form* (ed. J.T. Bonner, abridged edition of a book originally published in 1917, with second edition in 1942), Cambridge University Press, Cambridge.

[14]   Gánti, T. (2003) *The Principles of Life*, Oxford University Press, Oxford.

[15]   Maturana, H.R. and Varela, F.J. (1980) *Autopoiesis and Cognition: the Realisation of the Living*, D. Reidel Publishing Company, Dordrecht, The Netherlands.

[16]   Kauffman, S.A. (1986) Autocatalytic sets of proteins. *J. Theor. Biol.* **119**:1 – 24.

[17]   Buchner, E. (1897) Alkoholische Gährung ohne Hefezellen. *Ber. Dt. Chem. Ges.* **30**:117 – 124.

[18]   Kacser, H. (1987) On parts and wholes in metabolism, in *The organization of cell metabolism* (ed. G.R. Welch & J.S. Clegg), pp. 327 – 337, Plenum, New York.

[19]   Cornish-Bowden, A., Cárdenas, M.L. (2007) Organizational invariance in (M,R)-systems. *Chem. Biodiversity* **4**:2396 – 2406.

[20]   Letelier, J.-C., Soto-Andrade, J., F. Guíñez Abarzúa, F., Cornish-Bowden, A., Cárdenas, M.L. (2006) Organizational invariance and metabolic closure: analysis in terms of (M,R)-systems. *J. Theor. Biol.* **238**:949 – 961.

[21]   Cornish-Bowden, A., Cárdenas, M.L., Letelier, J.-C., Soto-Andrade, J. (2007) Beyond reductionism: metabolic circularity as a guiding vision for a real biology of systems. *Proteomics* **7**:839 – 845.

[22]   Meléndez-Hevia, E., Montero-Gómez, N, Montero, F. (2008) From prebiotic chemistry to cellular metabolism – the chemical evolution of metabolism before Darwinian natural selection. *J. Theor. Biol.* **252**:505 – 519.

[23]   Cottam, R., Ranson, W., Vounckx, R. (2007) Re-mapping Robert Rosen's (M,R)-systems. *Chem. Biodivers.* **4**:2352 – 2368.

[24]   Rosen, R. (1966) A note on replication in (M,R)-systems. *Bull. Math. Biophys.* **28**:149 – 151.

[25]   Rosen, R. (1971) Some realizations of (M,R)-systems and their interpretation. *Bull. Math. Biophys.* **33**:303 – 319.

[26]   Fell, D.A., Wagner, A. (2000) The small word of metabolism. *Nat. Biotechnol.* **18**:1121 – 1122.

[27]   Jeong, H., Tombor, B., Albert, R., Oltvai, Z.N., Barabási, A.L. (2000) The large-scale organization of metabolic networks. *Nature* **407**:651 – 654.

[28]   Csete, M. & Doyle, J. (2004) Bow ties, metabolism and disease. *Trends Biotechnol.* **22**:446 – 450.

[29]   Morán, F., Moreno, A., Minch, E., Montero, F. (1996) Further steps towards a realistic description of the essence of life. In Langton, C. G., Shimohara, K., editors, *Artificial Life V*, pp. 255–263. MIT Press, Cambridge, Massachusetts.

[30]   Tipton, K.F., O'Sullivan, M.I., Davey, G.P., O'Sullivan, J. (2003) It can be a complicated life being an enzyme. *Biochem. Soc. Trans.* **31**:711–715.

[31]   Sriram, G., Martinez, J.A., McCabe, E.R.B., Liao, J.C., Dipple, K.M. (2005) Single-gene disorders: what role could moonlighting enzymes play? *Am. J. Hum. Genet.* **76**:911–924.

[32]   Cornish-Bowden, A., Cárdenas, M.L. (2008) Self-organization at the origin of life. *J. Theor. Biol.* **252**:411–418.

[33]   Woese, C.R. (2004) A new biology for a new century. *Microb. Molec. Biol. Rev.* **68**:173–186.

 Beilstein-Institut

Systems Chemistry, May 26th – 30th, 2008, Bozen, Italy

New Concepts for Catalysis

Corinna M. Reisinger, Subhas Chandra Pan and Benjamin List*

Max-Planck-Institut für Kohlenforschung, 45470 Mülheim an der Ruhr, Germany

E-Mail: *list@mpi-muelheim.mpg.de

Received: 16th January 2009 / Published: 16th March 2009

Abstract

Organocatalysis, the catalysis with low-molecular weight catalysts where a metal is not part of the catalytic principle, can be as efficient and selective as metal- or biocatalysis. Important discoveries in this area include novel Lewis base-catalyzed enantioselective processes and, more recently, simple Brønsted acid organocatalysts that rival the efficiency of traditional metal-based asymmetric Lewis acid-catalysts. Contributions to organocatalysis from our laboratories include several new and broadly useful concepts such as enamine catalysis and asymmetric counteranion directed catalysis. Our lab has discovered the proline-catalyzed direct asymmetric intermolecular aldol reaction and introduced several other organocatalytic reactions.

Introduction: Organocatalysis

When chemists make chiral compounds – molecules that behave like object and mirror image, such as amino acids, sugars, drugs, or nucleic acids – they like to use asymmetric catalysis, in which a chiral catalyst selectively accelerates the reaction that leads to one mirror-image isomer, also called enantiomer. For decades, the generally accepted view has been that there are two classes of efficient asymmetric catalysts: enzymes and synthetic metal complexes [1]. However, this view is currently being challenged, with purely organic catalysts emerging as a third class of powerful asymmetric catalysts (Figure 1).

Most biological molecules are chiral and are synthesized in living cells by enzymes using asymmetric catalysis. Chemists also use enzymes or even whole cells to synthesize chiral compounds and for a long time, the perfect enantioselectivities observed in enzymatic reactions were considered beyond reach for non-biological catalysts. Such biological catalysis is increasingly used on an industrial scale and is particularly favoured for hydrolytic reactions. However, it became evident that high levels of enantioselectivity can also be achieved using synthetic metal complexes as catalysts. Transition metal catalysts are particularly useful for asymmetric hydrogenations, but may leave possibly toxic traces of heavy metals in the product.

In contrast, in organocatalysis, a purely organic and metal-free small molecule is used to catalyze a chemical reaction. In addition to enriching chemistry with another useful strategy for catalysis, this approach has some important advantages. Small organic molecule catalysts are generally stable and fairly easy to design and synthesize. They are often based on nontoxic compounds, such as sugars, peptides, or even amino acids, and can easily be linked to a solid support, making them useful for industrial applications. However, the property of organocatalysts most attractive to organic chemists may be the simple fact that they are organic molecules. The interest in this field has increased spectacularly in the last few years [2].

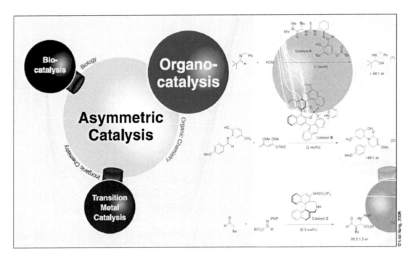

Figure 1. The three pillars of asymmetric catalysis: Biocatalyis, Metal Catalysis and Organocatalysis

Organocatalysts can be broadly classified as Lewis bases, Lewis acids, Brønsted bases, and Brønsted acids [3]. The corresponding (simplified) catalytic cycles are shown in Scheme 1. Accordingly, Lewis base catalysts (B:) initiate the catalytic cycle via nucleophilic addition to the substrate (S). The resulting complex undergoes a reaction and then releases the product (P) and the catalyst for further turnover. Lewis acid catalysts (A) activate nucleophilic substrates (S:) in a similar manner. Brønsted base and acid catalytic cycles are initiated via a (partial) deprotonation or protonation, respectively.

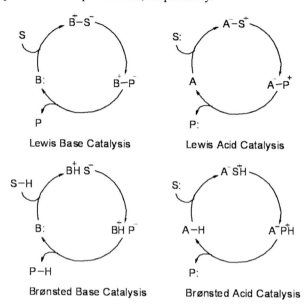

Scheme 1. Organocatalytic cycles.

ENAMINE CATALYSIS

Enamine catalysis involves a catalytically generated enamine intermediate that is formed via deprotonation of an iminium ion and that reacts with various electrophiles or undergoes pericyclic reactions. The first example of asymmetric enamine catalysis is the Hajos-Parrish-Eder-Sauer-Wiechert reaction [4] (Scheme 2), an intramolecular aldol reaction catalyzed by proline. Despite its use in natural product and steroid synthesis, the scope of the the Hajos-Parrish-Eder-Sauer-Wiechert reaction had not been explored, its mechanism was poorly understood, and its use was limited to a narrow context. Inspired by the development of elegant biocatalytic and transition metal complex-catalyzed direct asymmetric aldolizations [5], a revival of this chemistry was initiated with the discovery of the proline-catalyzed direct asymmetric intermolecular aldol reaction about thirty years later [6]. Since then proline-catalyzed enantioselective intermolecular aldol reactions [7], Mannich reactions [8] and Michael additions [9] have been developed [10].

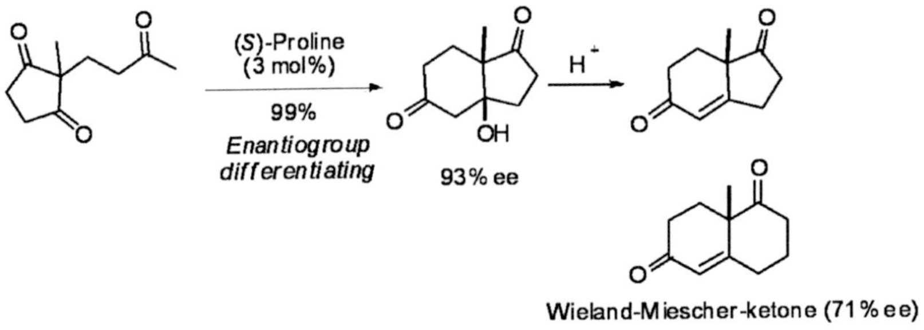

Wieland-Miescher-ketone (71% ee)

Scheme 2. The Hajos-Parrish-Eder-Sauer-Wiechert reaction.

This concept has also been extended to highly enantioselective α-functionalizations of aldehydes and ketones such as aminations [11], hydroxylations [12], alkylations [13], chlorination [14], fluorination [15], bromination [16], sulfenylation [17] and an intramolecular Michael reaction [18] using proline, as well as other chiral secondary amines and chiral imidazolidinones as the catalysts.

The proline-catalyzed asymmetric aldol reaction: scope, mechanism and consequences

In addition to catalyzing the well-known Hajos-Parrish-Eder-Sauer-Wiechert reaction (Scheme 3, eq. 1), we found in early 2000 that proline also catalyzes intermolecular aldolizations (e. g. eq. 2). Thereafter, our reaction has been extended to other substrate combinations (aldehyde to aldehyde, aldehyde to ketone, and ketone to ketone, eq. 3 – 5) and to enolexo-aldolizations (eq. 6) [7a, 7b, 19]. Proline seems to be a fairly general, efficient, and enantioselective catalyst of the aldol reaction and the substrate scope is still increasing continuously.

Both experimental and theoretical studies have contributed significantly to the elucidation of the reaction mechanism. We found that in contrast to earlier proposals [20], proline catalyzed aldol reactions do not show any non-linear effects in the asymmetric catalysis [21]. These lessons as well as isotope incorporation studies provided experimental support for our previously proposed single proline enamine mechanism and for Houk's similar DFT-model of the transition state of the intramolecular aldol reaction [22]. On the basis of these results we proposed the mechanism shown in Scheme 4. Key intermediates are the iminium ion and the enamine. Iminium ion formation effectively lowers the LUMO energy of the system. As a result, both nucleophilic additions and α-deprotonation become more facile. Deprotonation leads to the generation of the enamine, which is the actual nucleophilic carbanion equivalent. Its reaction with the aldehyde then provides, via transition state **TS** and hydrolysis, the enantiomerically enriched aldol product.

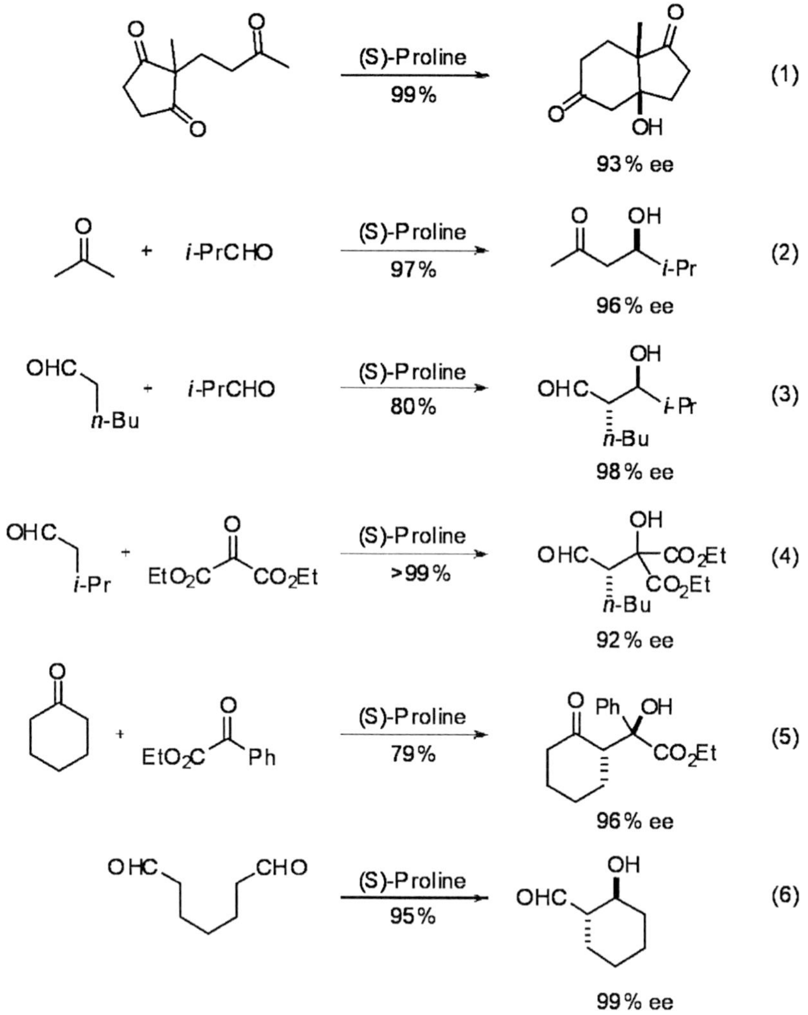

(S)-Proline
99%
(1)
93% ee

+ i-PrCHO
(S)-Proline
97%
(2)
96% ee

OHC
n-Bu
+ i-PrCHO
(S)-Proline
80%
(3)
OHC
n-Bu
i-Pr
OH
98% ee

OHC
i-Pr
+ EtO₂C CO₂Et
(S)-Proline
>99%
(4)
OHC
n-Bu
CO₂Et
CO₂Et
OH
92% ee

+ EtO₂C Ph
(S)-Proline
79%
(5)
Ph OH
CO₂Et
96% ee

OHC CHO
(S)-Proline
95%
(6)
OHC
OH
99% ee

Scheme 3. Proline-catalyzed asymmetric aldol reactions.

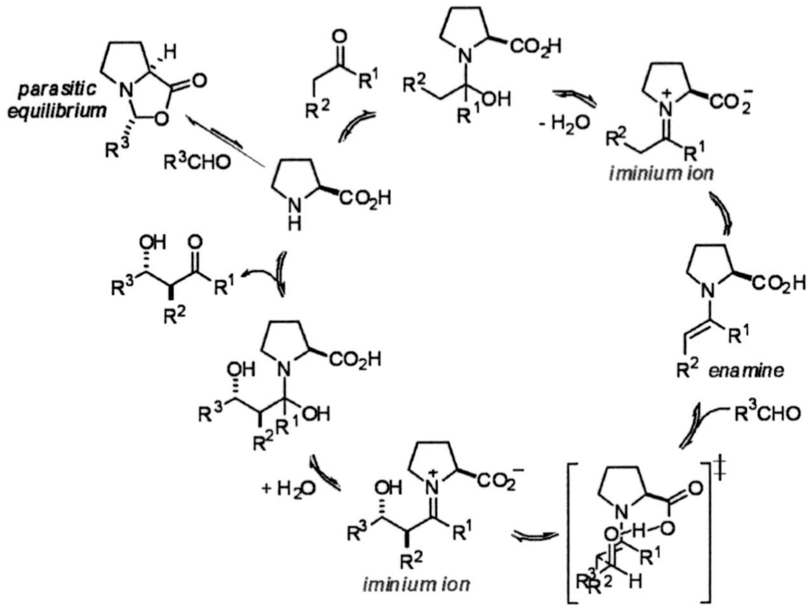

Scheme 4. Proposed mechanism and transition state of proline-catalyzed aldolizations.

For us, the intriguing prospect arose, that the catalytic principle of the proline-catalyzed aldol reaction may be far more general than originally thought. We reasoned that simple chiral amines including proline should be able to catalytically generate chiral enamines as carbanion equivalents, which then may undergo reactions with various electrophiles. We termed this catalytic principle *enamine catalysis* (Scheme 5) [23]. Accordingly, the enamine, which is generated from the carbonyl compound via iminium ion formation can react with an electrophile X=Y (or X-Y) via nucleophilic addition (or substitution) to give an α-modified iminium ion and upon hydrolysis the α-modified carbonyl product (and HY).

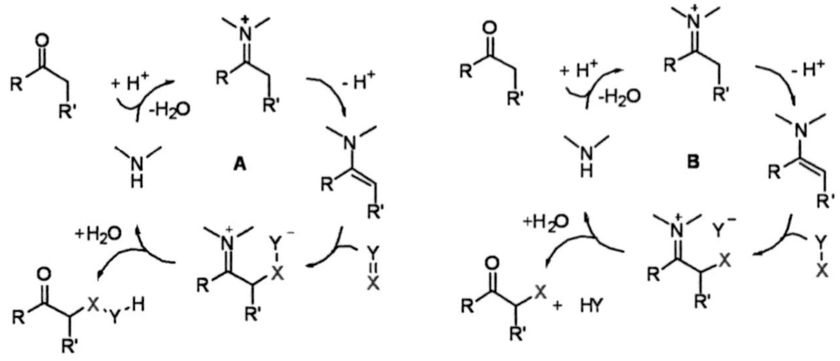

Scheme 5. Enamine catalysis of nucleophilic addition (**A**)- and substitution (**B**) reactions (arrows may be considered equilibria).

Enamine catalysis has developed dramatically in the last few years and it turns out that its scope not only exceeds our most optimistic expectations but also that of the traditional stoichiometric enamine chemistry of Stork and others.

Enamine catalysis of nucleophilic addition reactions

Enamine catalysis using proline or related catalysts has now been applied to both intermolecular and intramolecular nucleophilic addition reactions with a variety of electrophiles. In addition to carbonyl compounds (C=O), these include imines (C=N) in Mannich reactions [8], azodicarboxylates (N=N) [11], nitrosobenzene (O=N) [12], and Michael acceptors (C=C) [18, 24] (see Scheme 6, eq. 7–10 for selected examples).

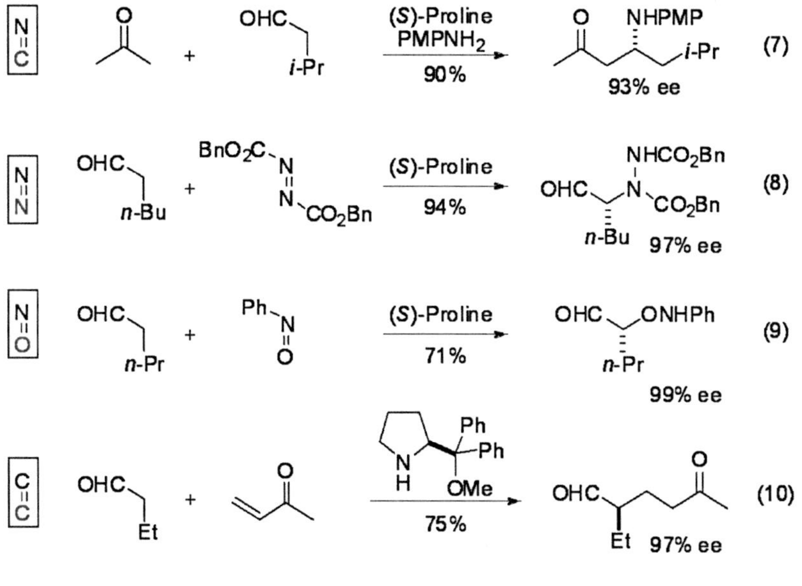

Scheme 6. Enamine catalysis of nucleophilic addition reactions.

Enamine catalysis often delivers valuable chiral compounds such as alcohols, amines, aldehydes, and ketones. Many of these are normally not accessible using established reactions based on transition metal catalysts or on preformed enolates or enamines, illustrating the complimentary nature of organocatalysis and metallocatalysis.

Enamine catalysis of nucleophilic substitution reactions

The first example of an asymmetric enamine catalytic nucleophilic substitution was a reaction that may have been considered impossible only a few years ago. We found that proline and certain derivatives such as α-methyl proline efficiently catalyze the asymmetric α-alkylation of aldehydes [13]. Catalytic α-alkylation reactions of substrates other than glycine derivatives have been rare and that of aldehydes has been completely unknown before. In our process we could cyclize 6-halo aldehydes to give cyclopentane carbaldehydes in excellent yields and *ee*s (Scheme 7, eq. 11). Other important and remarkably useful enamine catalytic nucleophilic substitution reactions have been developed subsequently and include enantioselective α-chlorinations [14], α-fluorinations [15], α-brominations [16], α-iodinations, and α-sulfenylations [17] (eq. 12 – 16).

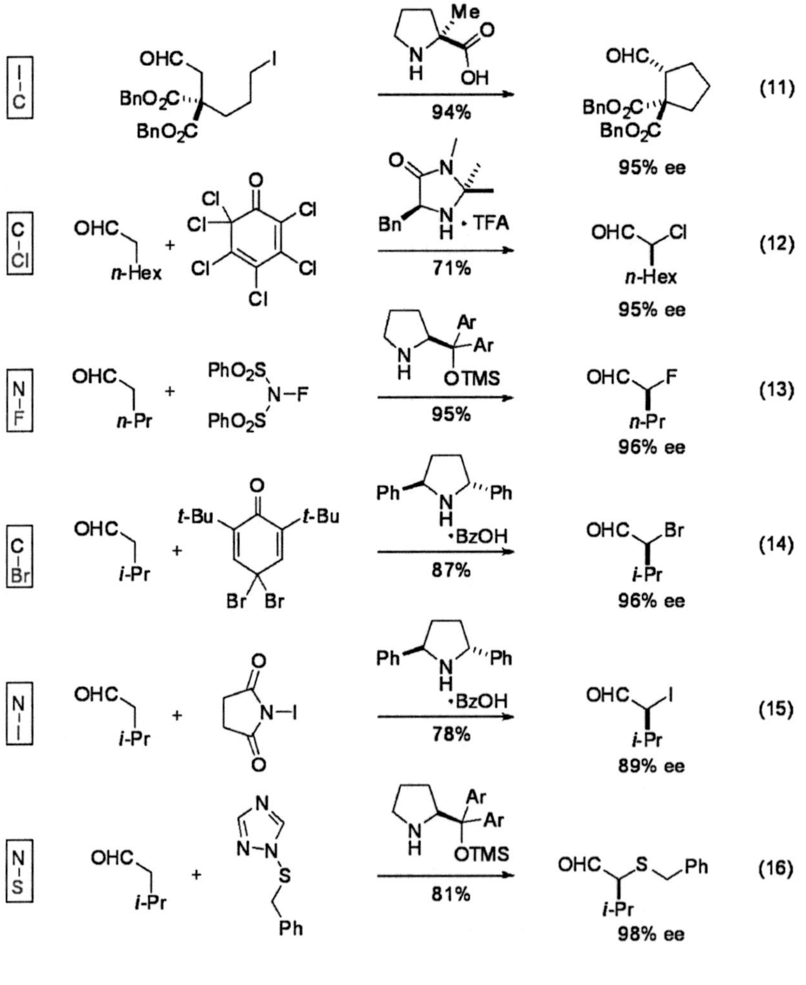

Scheme 7. Enamine catalysis of nucleophilic substitution reactions.

Once again, most of these reactions have never been realized before using preformed enamines or any other methodology but lead to highly valuable products of potential industrial relevance.

The proline-catalyzed asymmetric Mannich reactions

The catalytic asymmetric Mannich reaction is arguably the most useful approach to chiral α-amino carbonyl compounds. In the year 2000, we discovered a proline-catalyzed version of this powerful reaction [8a]. Originally, ketones, aldehydes, and an aniline as the amine component were used in a catalytic asymmetric three-component reaction (Scheme 8, eq 17). After our report, proline catalyzed Mannich reactions with aldehydes as the donor have also been developed [8d, 8e] (eq 18 – 19). Despite its frequent use, both in an academic as well as an industrial context, the main limitation of the proline-catalyzed Mannich reaction has been the requirement to use anilines as the amine component. Although optically enriched p-anisidylamines are of potential utility in asymmetric synthesis, facile and efficient removal of the N-protecting group to yield the unfunctionalized amine is required. Generally, the removal of the most commonly used p-methoxyphenyl (PMP) group from nitrogen requires rather drastic oxidative conditions involving harmful reagents such as ceric ammonium nitrate that are not compatible with all substrates. We have now identified reaction conditions that allow for the use of simple preformed aromatic N-Boc-imines in proline-catalyzed Mannich reactions (eq. 20). Remarkably, the reaction provides chiral β-amino aldehydes and ketones as stable, crystalline compounds in generally high diastereo- and enantioselectivities without the requirement for chromatographic purification [25].

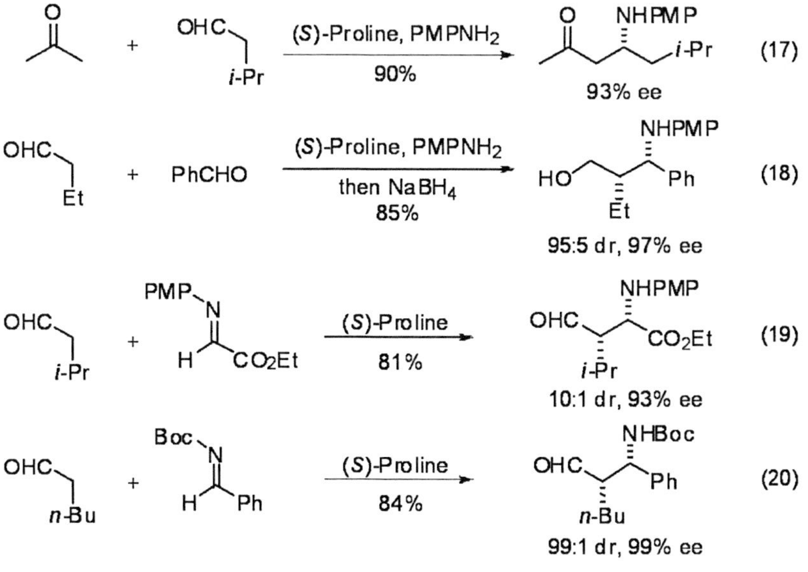

Scheme 8. Proline-catalyzed asymmetric Mannich reactions.

A typical experimental procedure is illustrated in Figure 2. Mixing the 2-naphthaldehyde-derived *N*-Boc imine with isovaleraldehyde in the presence of (S)-proline (20 mol%) in acetonitrile at 0 °C resulted in an initially homogenous reaction mixture (Figure 2a). After complete consumption of the starting material (10 h), a large amount of the desired product had precipitated and could easily be collected by filtration (Figure 2b).

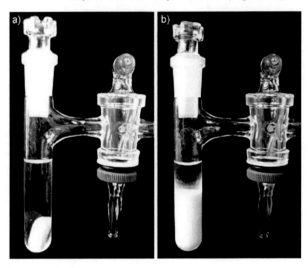

Figure 2. The reaction of isovaleraldehyde with 2-naphthyl *N*-Boc-imine in the presence of (S)-proline (20 mol%) in CH₃CN. (**a**) Homogenous reaction mixture after mixing all components. (**b**) Reaction mixture after completion of the reaction (10 h).

The *N*-Boc-imine-derived Mannich products can readily be converted into the corresponding α,β-branched-β-amino acids (β^{2,3}-amino acids). For example, oxidation of the product **1** to the carboxylic acid followed by acid-mediated deprotection provided the amino acid salt **2** without loss of stereochemical integrity (Scheme 9, TFA = trifluoroacetic acid). Measuring NMR spectra and optical rotation of the corresponding HCl salt allowed us to confirm the expected absolute and relative configuration of the product.

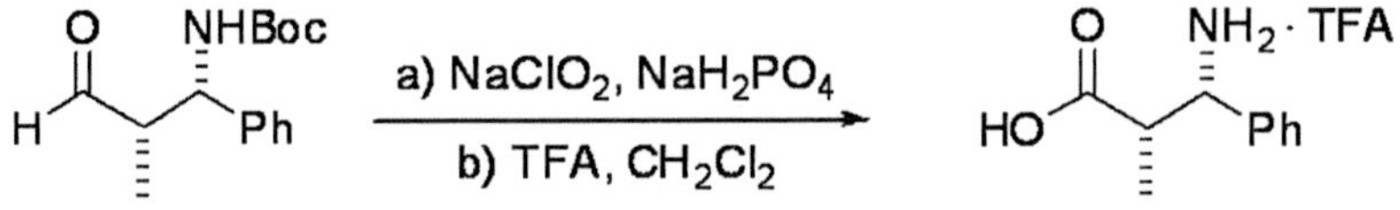

Scheme 9. Conversion of the Mannich product **1** to the β-amino acid **2**.

BRØNSTED ACID CATALYSIS

In the proline-based enamine catalysis, proline actually plays a dual role. The amino-group of proline acts as a Lewis base, whereas the carboxylic group acts as a Brønsted acid (Scheme 10).

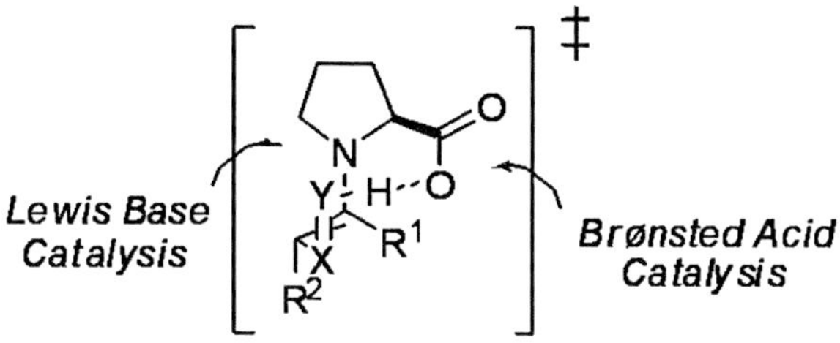

Scheme 10. Proline: a bifunctional catalyst.

The potential of using relatively strong chiral organic Brønsted acids as catalysts (Specific Brønsted acid catalysis) has been essentially ignored over the last decades. Achiral acids such as p-TsOH have been used as catalysts for a variety of reactions since a long time, but applications in asymmetric catalysis have been extremely rare. Only very recently, Akiyama *et al.* [26] and Terada *et al.* [27] in pioneering studies demonstrated that relatively strong chiral binaphthol-derived phosphoric acids are efficient and highly enantioselective catalysts for addition reactions to aldimines (Scheme 11).

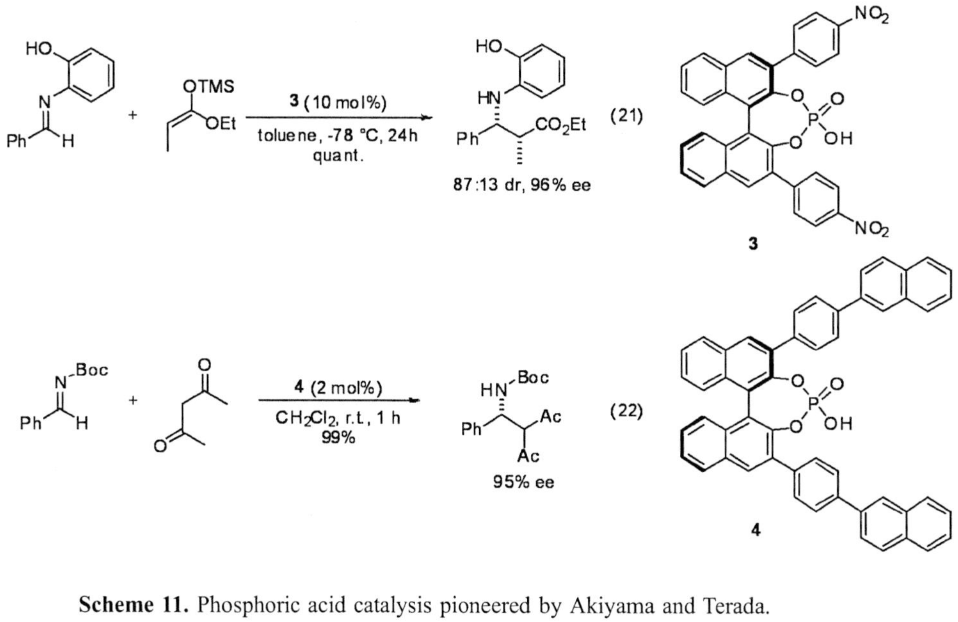

Scheme 11. Phosphoric acid catalysis pioneered by Akiyama and Terada.

Catalytic asymmetric Pictet-Spengler reaction

The Pictet-Spengler reaction [28] is an important acid-catalyzed transformation frequently used in the laboratory as well as by various organisms for the synthesis of tetrahydro-β-carbolines and tetrahydroisoquinolines from carbonyl compounds and 2-phenylethylamines or tryptamines, respectively.

Very recently, Jacobsen *et al.* [29] reported the first truly catalytic version by using an elegant organocatalytic acyl-Pictet-Spengler approach. The direct Pictet-Spengler reaction of aldehydes with 2-arylethylamines however, has been an illusive target for small molecule catalysis. Since the addition reactions to aldimines developed by Akiyama and Terada are assumed to involve chiral iminium phosphate ion pairs, we reasoned that a chiral phosphoric acid-catalyzed approach might be as well applicable to the Pictet-Spengler reaction, which also proceeds via iminium ion intermediates.

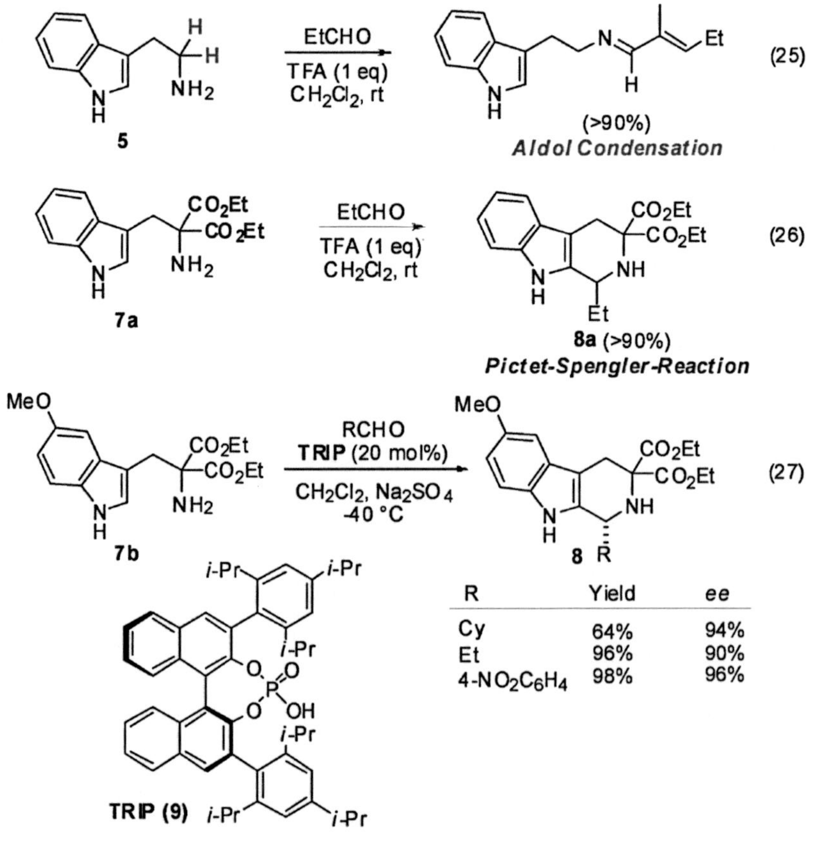

R	Yield	ee
Cy	64%	94%
Et	96%	90%
4-NO$_2$C$_6$H$_4$	98%	96%

Scheme 12. Brønsted acid-catalyzed Pictet-Spengler reaction.

In line with observations by Jacobsen *et al.* attempts toward Brønsted acid catalysis of the Pictet-Spengler reaction of simple substrates such as unsubstituted tryptamines **5** and 2-phenylethylamines failed due to competing homoaldol condensation followed by imine formation (Scheme 12, eq. 25). A solution to this problem was the use of more reactive substrates such as geminally disubstituted tryptamines **7a** [30] predisposed for cyclization by virtue of a Thorpe-Ingold effect. Treatment of **7a** with TFA cleanly provided the desired Pictet-Spengler product **8a** in > 90 % yield (eq. 26). Encouraged by this result we went on to develop an asymmetric version with the use of a chiral Brønsted acid catalyst. In the presence of BINOL phosphate **TRIP (9)** bearing bulky 2,4,6-triisopropylphenyl substituents at the 3,3'-positions of the binaphthyl scaffold and Na_2SO_4, tetrahydro-β-carbolines **8** were obtained in high yields along with excellent enantioselectivities (eq. 27) [31]. Remarkably, the reaction tolerates a variety of both aliphatic and aromatic aldehydes with excellent results.

Organocatalytic asymmetric reductive amination

Catalytic asymmetric hydrogenations are among the most important transformations in organic chemistry. Although numerous methods employing olefins or ketones as substrates have been described [32], the corresponding hydrogenations or transfer hydrogenations of imines are less advanced [33]. Living organisms apply cofactors such as nicotinamide adenine dinucleotide (NADH) for enzyme-catalyzed reductions of imines [34].

Inspired by the recent observation that imines are reduced with Hantzsch dihydropyridines as a NADH analogue in the presence of achiral Lewis or Brønsted acid catalysts [35], we envisioned a catalytic cycle for the reductive amination of ketones which is initiated by protonation of the *in situ* generated ketimine **10** by a chiral Brønsted acid catalyst (Scheme 13). The resulting iminium ion pair is chiral and its reaction with the Hantzsch ester **11** could give rise to enantiomerically enriched α-branched amine **12** and pyridine **13**.

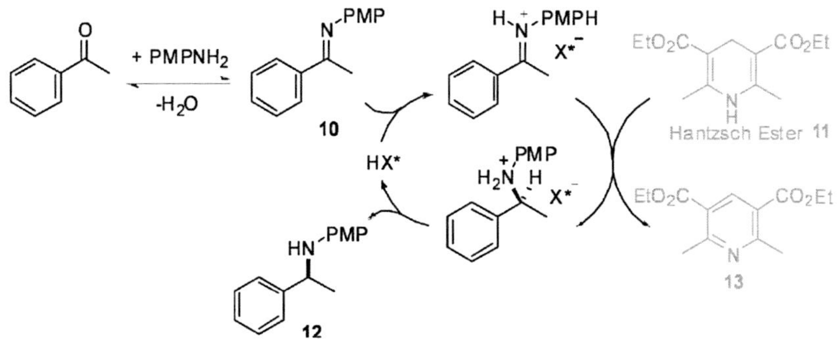

Scheme 13. Chiral Brønsted acid-catalyzed reductive amination.

Among all the phosphoric acids tested as chiral Brønsted acid catalyst in this reaction, TRIP was found to be the best. Only 1 mol% of TRIP was sufficient to give the desired product in an excellent yield of 96% and with 93% ee. (Scheme 14, eq. 29) [36]. A similar study by the Rueping group using Akiyama's phosphoric acid catalyst **14** appeared during the preparation of our manuscript (eq. 30) [37]. MacMillan and co-workers also developed a reductive amination of various ketones catalyzed by BINOL phosphate **15** (eq. 31) [38].

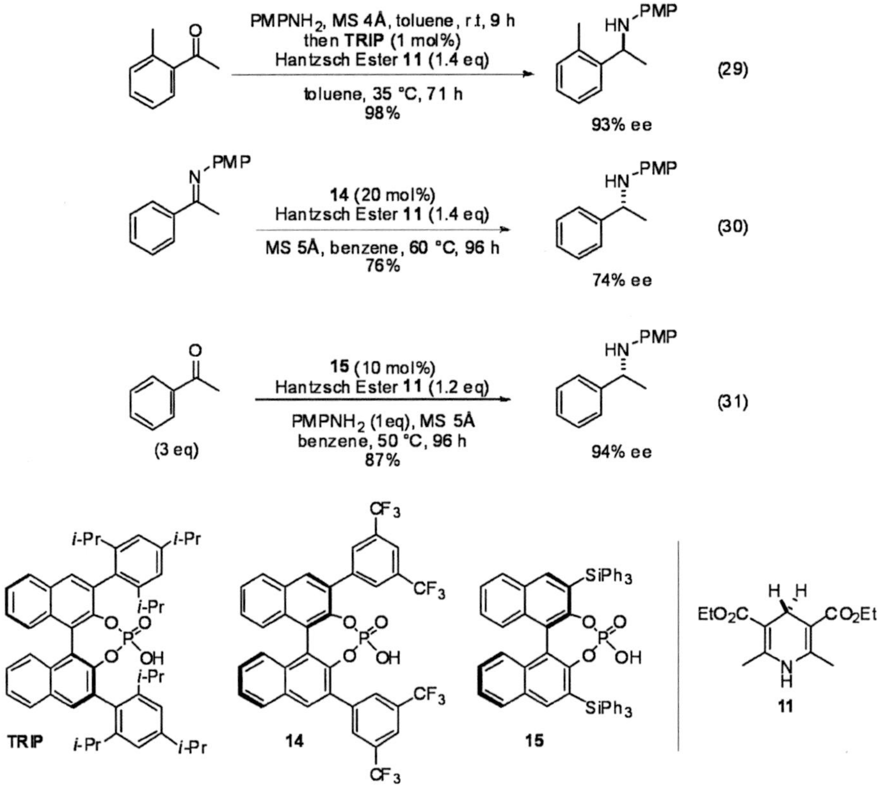

Scheme 14. Chiral phosphoric acid-catalyzed asymmetric reductive amination of ketones.

The previous examples are selected asymmetric reductive aminations of ketones to give chiral, α-branched amines (eq. 32); however, the corresponding reactions of aldehydes are unknown. We reasoned that such a process might be realized if enolizable, α-branched aldehydes are employed. Their asymmetric reductive amination should give β-branched amines via an enantiomer-differentiating kinetic resolution (eq. 33).

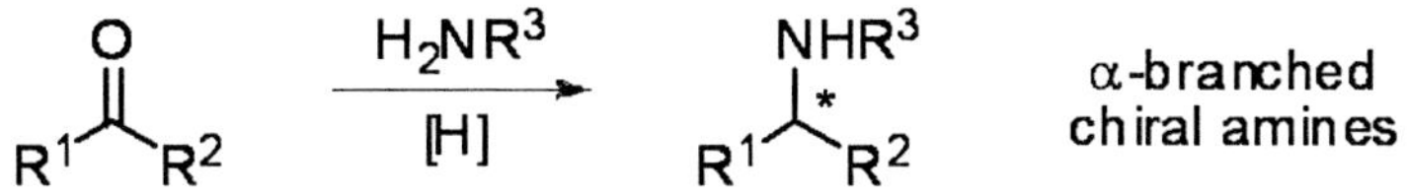

$$\text{(32)} \quad \alpha\text{-branched chiral amines}$$

R^{1}COR^2 + H$_2$NR^3 / [H] → R^{1}C(NHR^3)R^2 (*) — α-branched chiral amines (32)

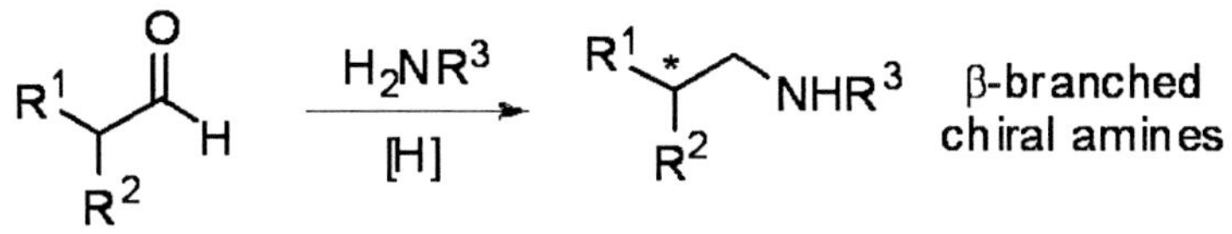

R^1(R^2)CHCHO + H$_2$NR^3 / [H] → R^1(*)CH(R^2)CH$_2$NHR^3 — β-branched chiral amines (33)

At the onset of this study, we hypothesized that under our reductive amination conditions an α-branched aldehyde substrate would undergo a fast racemization in the presence of the amine and acid catalyst via an imine/enamine tautomerization. The reductive amination of one of the two imine enantiomers would then have to be faster than that of the other, resulting in an enantiomerically enriched product via a dynamic kinetic resolution (Scheme 15) [39].

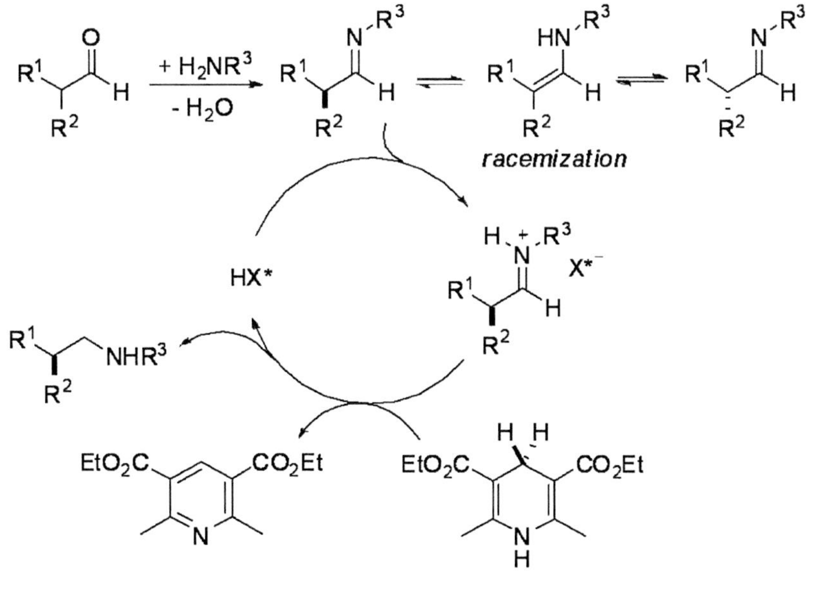

Scheme 15. Catalytic asymmetric reductive amination of aldehydes.

Indeed, when we studied various phosphoric acid catalysts for the reductive amination of hydratopicaldehyde (**16**) with *p*-anisidine (PMPNH$_2$) in the presence of Hantzsch ester **11** to give amine **17**, the observed enantioselectivities and conversions are consistent with a facile *in situ* racemization of the substrate and a resulting dynamic kinetic resolution (Scheme 16).

TRIP (**9**) once again turned out to be the most effective and enantioselective catalyst for this transformation and provided the chiral amine products with different α-branched aldehydes and amines in high enantioselectivities [40].

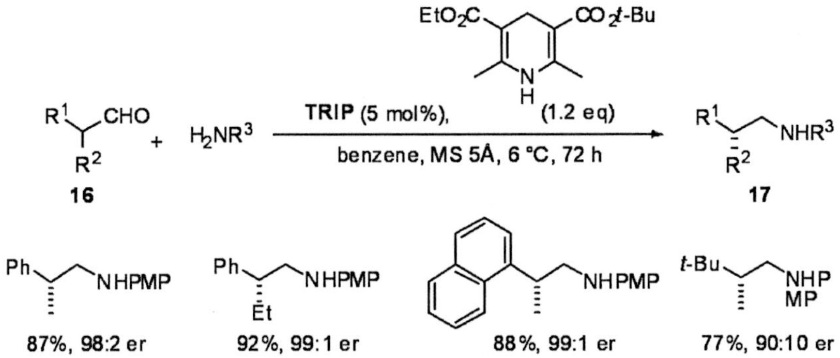

87%, 98:2 er   92%, 99:1 er   88%, 99:1 er   77%, 90:10 er

Scheme 16. Catalytic asymmetric reductive amination of aldehydes using **TRIP**.

We later developed an analogous enantioselective hydrogenation of aldehydes to the corresponding β-branched alcohols using [RuCl$_2$(xylyl-BINAP)(DPEN or DACH)] as the catalyst [41].

IMINIUM CATALYSIS

The *in situ* generation of an iminium ion from a carbonyl compound lowers the LUMO energy of the system. *Iminium catalysis* is comparable to Brønsted- or Lewis acid activation of carbonyl compounds. The LUMO energy is lowered, the α-CH-acidity increases, and nucleophilic additions including conjugate additions as well as pericyclic reactions are facilitated (eq. 34).

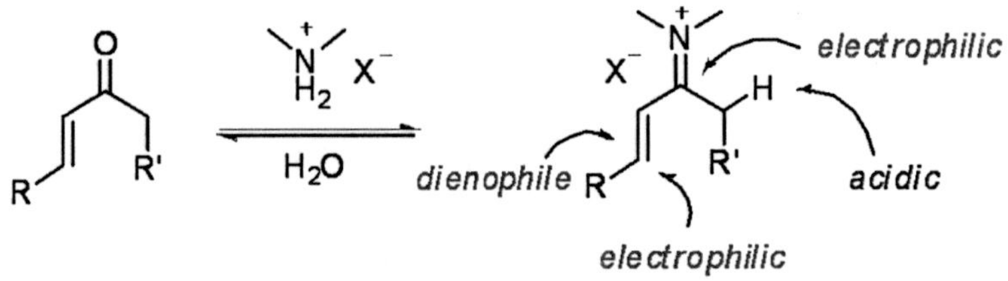

(34)

The first highly enantioselective examples of this catalysis strategy were reported by Mac-Millan *et al.* in 2000 [42], shortly after our first report on the proline-catalyzed intermolecular aldol reaction had appeared. The MacMillan group has quickly established that Diels-Alder reactions, 1,3-dipolar cycloadditions [43], and conjugate additions of electron rich aromatic and heteroaromatic compounds can be catalyzed using chiral amino acid derived imidazolidinones as catalysts (Scheme 17, eq. 35–38) [44]. In addition, highly enantioselective epoxidations [45] and cyclopropanations [46] have recently been developed.

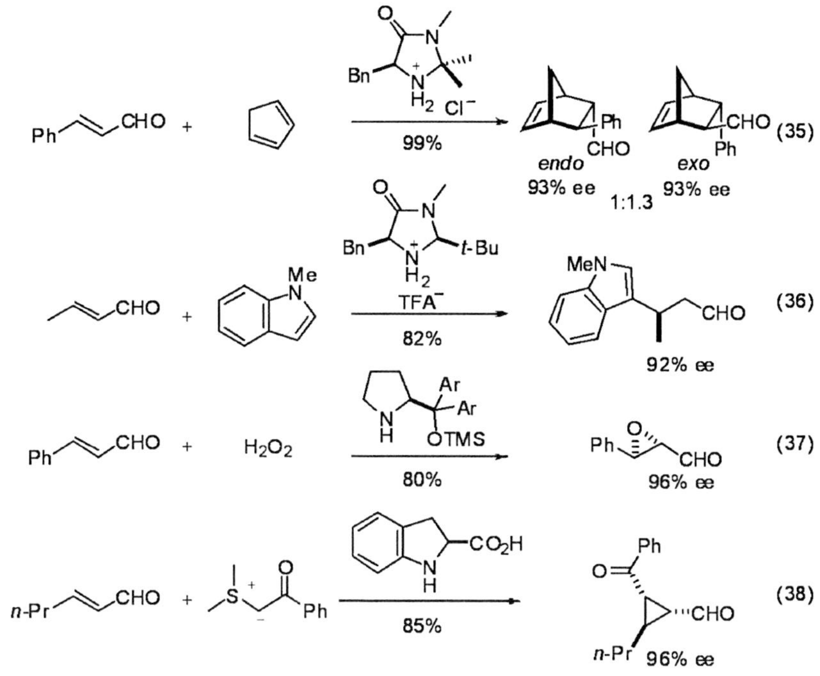

Scheme 17. Iminium catalytic asymmetric transformations.

Organocatalytic conjugate reduction of α,β-unsaturated aldehydes

In 2001, we reasoned that this catalysis strategy might be applicable to the conjugate reduction of α,β-unsaturated carbonyl compounds if a suitable hydride-donor could be identified. Hantzsch ester **11** was chosen as the hydride source for this reaction (Scheme 18).

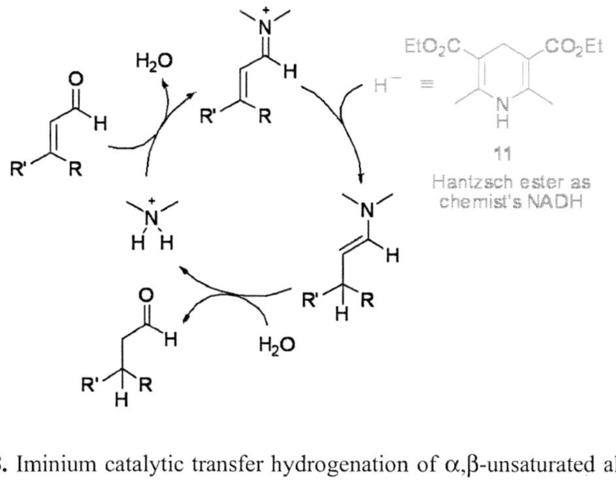

Scheme 18. Iminium catalytic transfer hydrogenation of α,β-unsaturated aldehydes.

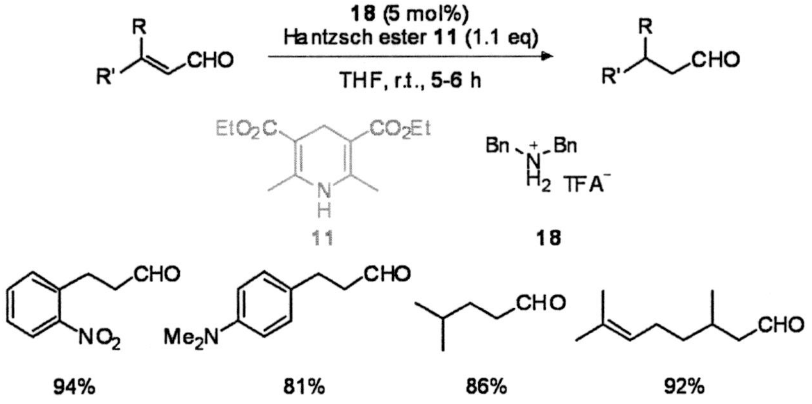

Scheme 19. Organocatalytic transfer hydrogenation of enals.

This process was published in 2004 and constitutes the first metal-free organocatalytic transfer hydrogenation of α,β-unsaturated aldehydes [47a]. Dibenzylammonium trifluoroacetate **18**, was found to be an efficient catalyst for this reaction. The reduction worked extremely well with a diverse set of unsaturated aldehydes, including substituted aromatic and aliphatic ones and the yields exceed 90% in almost all cases (Scheme 19). A variety of functional groups that are sensitive to standard hydrogenation condition (nitro, nitrile, benzyloxy, and alkene functional groups) were tolerated in the process.

The first example of an asymmetric catalytic version was also presented in our first publication [47a]. This protocol was subsequently optimized and we developed a highly enantioselective variant using the trichloroacetate salt of MacMillan's second generation imidazolidinone (**19**) as the catalyst [47b]. We found that upon treating aromatic, trisubstituted α,β-unsaturated aldehydes **20** with a slight excess of dihydropyridine **21** and a catalytic amount of **19** at 13 °C in dioxane, the corresponding saturated aldehydes **22** were obtained in high yields and enantioselectivities (Scheme 20).

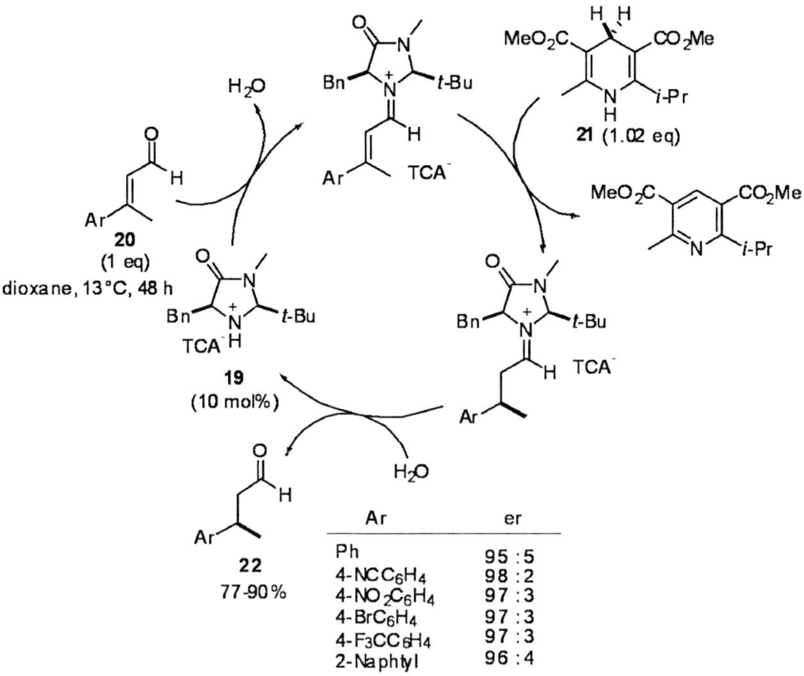

Scheme 20. Organocatalytic asymmetric transfer hydrogenation of enals.

Ar	er
Ph	95 : 5
4-NC C$_6$H$_4$	98 : 2
4-NO$_2$C$_6$H$_4$	97 : 3
4-BrC$_6$H$_4$	97 : 3
4-F$_3$CC$_6$H$_4$	97 : 3
2-Naphtyl	96 : 4

ASYMMETRIC COUNTERANION-DIRECTED CATALYSIS (ACDC)

Most chemical reactions proceed via charged intermediates or transition states. In asymmetric Brønsted acid catalysis the substrate is protonated by the catalyst and a chiral H-bond-assisted ion pair is generated. We reasoned that in principle any reaction that proceed via cationic intermediates can be conducted highly enantioselectively if a chiral counteranion is introduced into the catalyst, as a result of the generation of a chiral ion pair. We termed this new strategy as *Asymmetric Counteranion-Directed Catalysis* (ACDC) (Scheme 21).

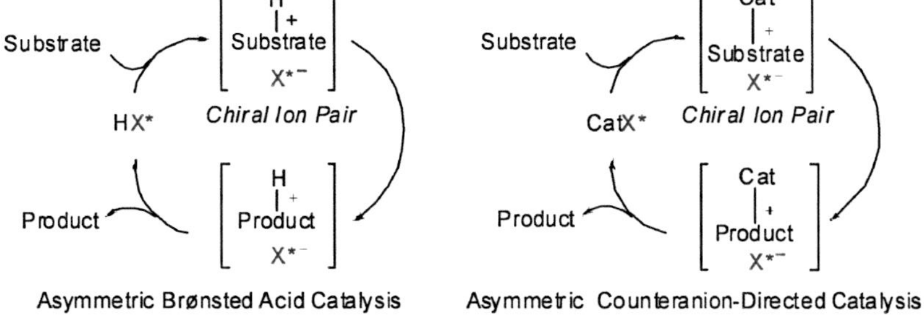

Scheme 21. Asymmetric counteranion-directed catalysis (ACDC).

Although efficient asymmetric catalytic transformations involving anionic intermediates with chiral, cationic catalysts have been realized [48], analogous versions of inverted polarity with reasonable enantioselectivity, despite attempts, have been illusive [49].

Asymmetric counteranion-directed catalysis: application to iminium catalysis

In iminium catalysis, both we and the group of MacMillan had observed a strong counteranion effect on the yield and enantioselectivity of the reactions. Inspired by recent use of chiral phosphoric acid derivatives as asymmetric catalysts, we hypothesized that catalytic salts of achiral amines and chiral phosphoric acids could induce asymmetry in these processes (Scheme 22).

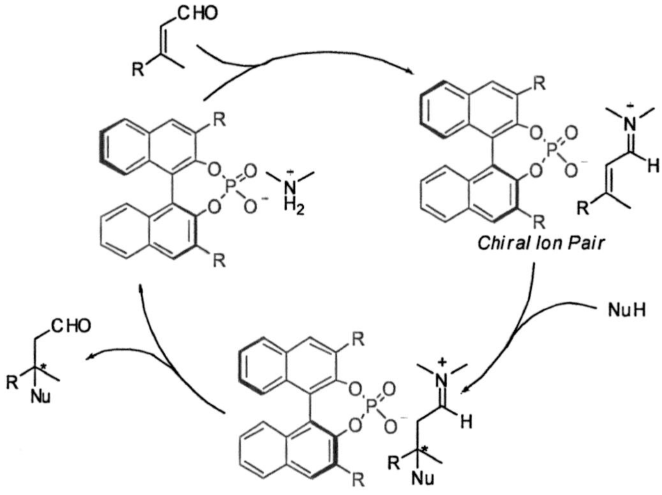

Scheme 22. Asymmetric counteranion-directed catalysis: application to iminium catalysis.

We thought to start with the metal-free biomimetic transfer hydrogenation of α,β-unsaturated aldehydes as a model reaction which has been earlier discovered in our laboratory and independently in that of MacMillan *et al.* (Scheme 23). We have prepared a large number of ammonium salts as crystalline solids by mixing different primary and secondary amines with a chiral phosphoric acid. In particular, the ammonium salts of sterically hindered chiral phosphoric acids could catalyze the reaction with significant enantiomeric excess (ee) values (Scheme 23). After a thorough screening of various amines we identified morpholine salt **28** as a highly enantioselective catalyst [50].

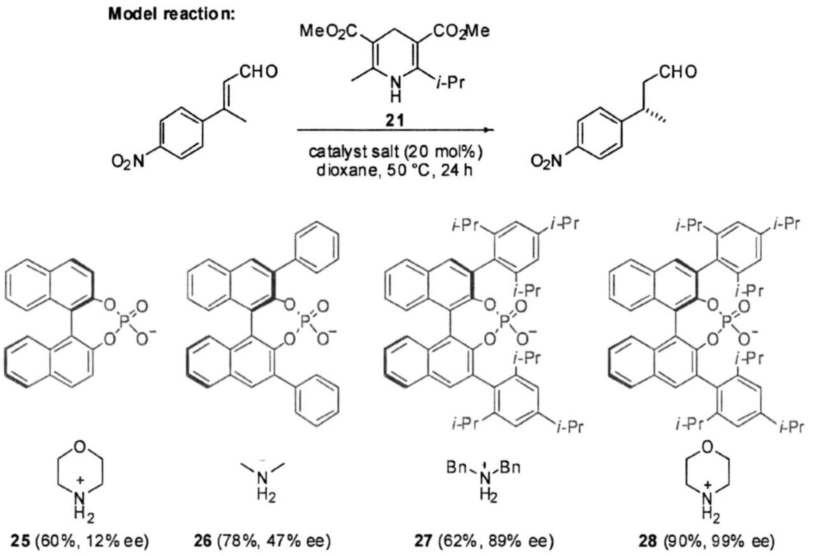

Model reaction:

Scheme 23. ACDC: Screening studies.

Treating aromatic, trisubstituted α,β-unsaturated aldehydes **20** with a slight excess of dihydropyridine **21** and a catalytic amount of salt **28** at 50 °C in dioxane for 24 h, the corresponding saturated aldehydes **22** were obtained in high yields and in enantioselectivities of 96–99% ee (Scheme 24).

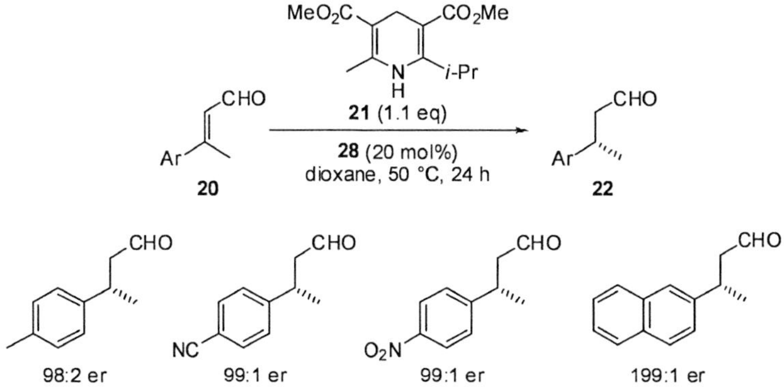

Scheme 24. ACDC: Transfer hydrogenation of enals.

Significantly, the previously developed chiral amine based catalysts that we and MacMillan and co-workers have studied have not been of use for sterically nonhindered aliphatic substrates. For example, citral (**29**), of which the hydrogenation product citronellal (**30**) is an intermediate in the industrial synthesis of menthol and used as a perfume ingredient, could not readily be used (Scheme 25, eq. 41). We could not achieve high enantioselectivity for this particular substrate with either our previous system [47b] or with that of MacMillan and coworkers [47c]. However, with our novel chiral counteranion catalyst **28**, citral is converted into (R)-citronellal (**30**) with an e.r. value of 95:5. This has been the highest enantioselectivity reported for a catalytic asymmetric (transfer) hydrogenation of citral [51]. Similarly, farnesal (**31**) gave (R)-dihydrofarnesal (**32**) in 77% yield and 96:4 er (Scheme 25, eq. 42).

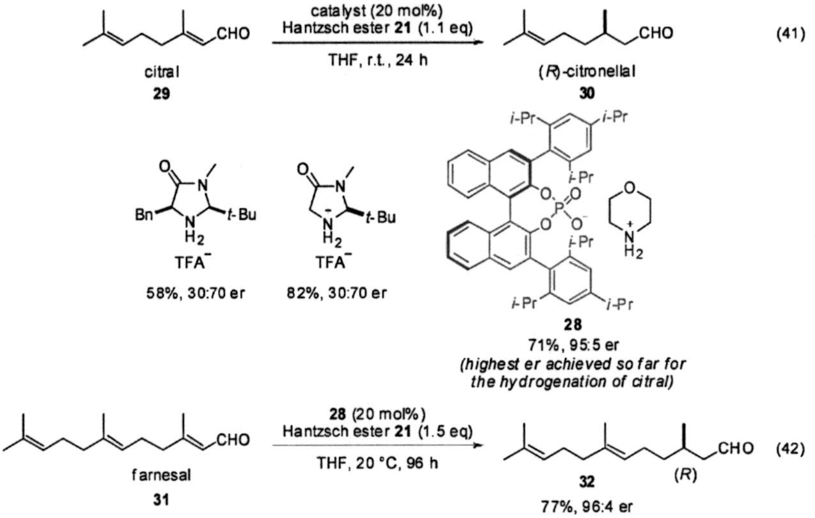

Scheme 25. ACDC: Transfer hydrogenation of citral and farnesal.

Next, we sought to extend this methodology to the conjugate reduction of α,β-unsaturated ketones. However, neither these ACDC-catalysts, nor the commonly used chiral imidazolidinone-catalysts gave satisfying yields or enantioselectivities in the conjugate reduction of 3-methyl cyclohexenone **33** (Scheme 26).

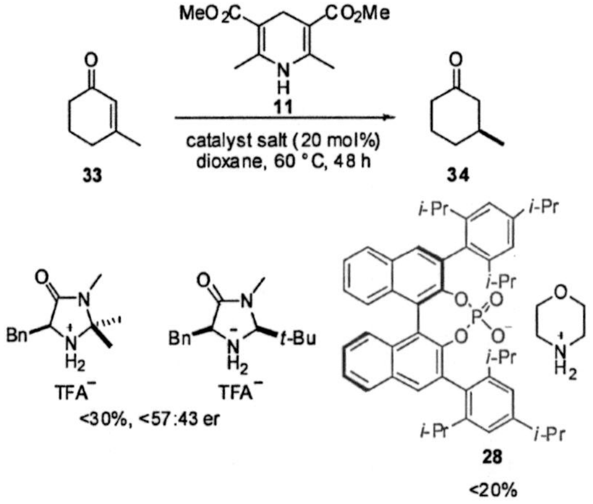

Scheme 26. ACDC: Transfer hydrogenation of 3-methyl cyclohexenone: first attempts.

Hypothesizing that primary amine salt catalysts might be suitable for the activation of ketones due to their reduced steric requirements we studied various salts of α-amino acid esters [52]. We have developed a new class of catalytic salts, in which both the cation and the anion are chiral. In particular, valine ester phosphate salt **35** proved to be an active catalyst for the transfer hydrogenation of a variety of α,β-unsaturated ketones **36** with commercially available Hantzsch ester **11** to give saturated ketones **37** in excellent enantios-electivities (Scheme 27) [53].

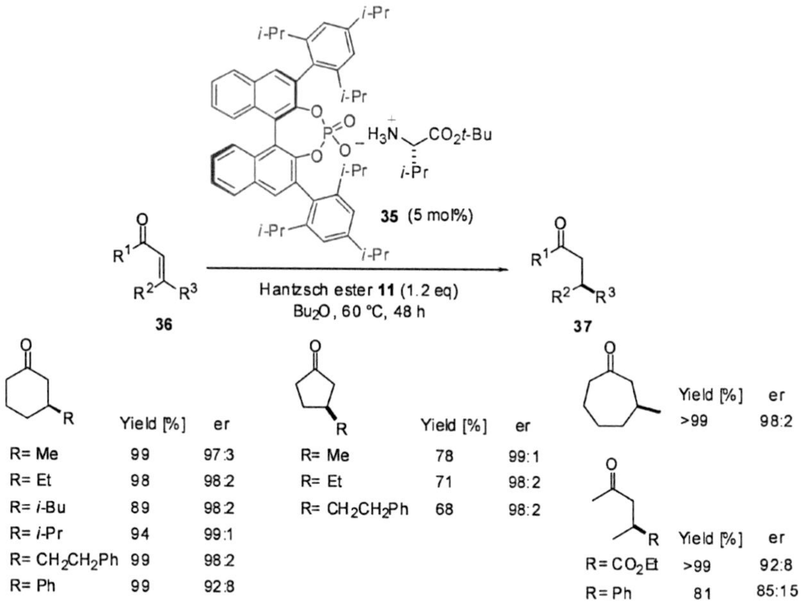

Scheme 27. ACDC: Transfer hydrogenation of enones.

Independently, MacMillan *et al.* developed an efficient catalyst system based on a chiral secondary amine for the transfer hydrogenation of cyclic enones [54].

Furthermore, the ACDC concept was successfully applied to the asymmetric epoxidation of enals (Scheme 28) [55]. Among all the combinations tested the catalyst salt **38** comprising an achiral dibenzylamine derivative together with a chiral binol phosphate counteranion proved to be the catalyst of choice furnishing the desired epoxides in good yields along with high enantioselectivities. Remarkably, also β,β-disubstituted α,β-unsaturated aldehydes gave the corresponding epoxides with excellent enantioselectivities in presence of the ACDC catalyst **38**. This is in sharp contrast to the results obtained when using the system described by Jørgensen and co-workers, where this substrate class could be converted into the corresponding epoxides only with moderate enantioselectivities [45].

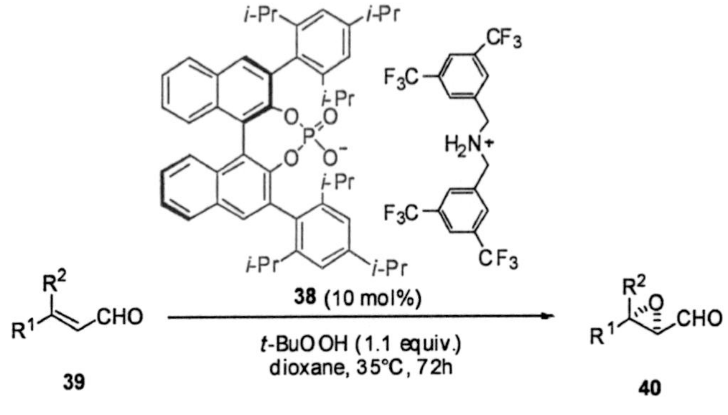

	Yield [%]	dr	er
R^1 = Ph	75	>99:1	95.5:4.5
R^1 = 4-MeC_6H_4	65	>99:1	96:4
R^1 = 3-MeC_6H_4	68	>99:1	96:4
R^1 = 2-MeC_6H_4	62	97:3	95.5:4.5
R^1 = 4-FC_6H_4	78	>99:1	97:3
R^1 = *n*-Hex	67	94:6	85:15

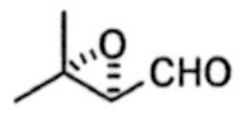

83%, 97:3 er

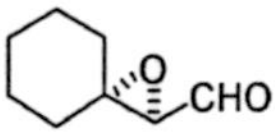

75%, 95:5 er

Scheme 28. ACDC: Epoxidation of enals.

The high enantioselectivity observed with these trisubstituted substrates raises interesting mechanistic questions. Since the initial addition product is achiral, the stereogenic center is created in the subsequent cyclization. Consequently, the chiral phosphate must be involved in this C-O bond-forming event and we propose the enantioselectivity to result from a **TRIP**-assisted cyclization of the achiral enamine intermediate (Scheme 29).

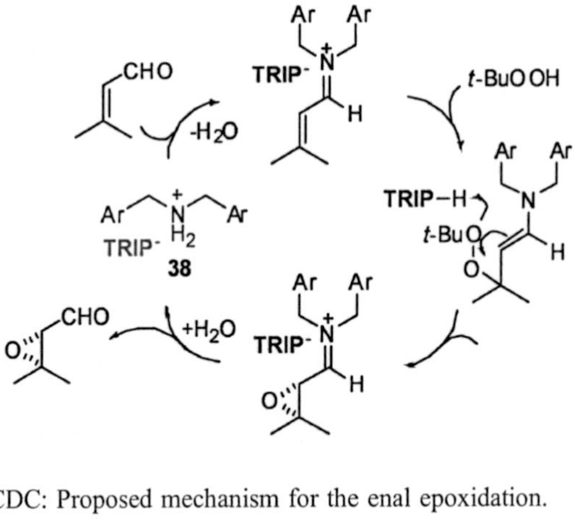

Scheme 29. ACDC: Proposed mechanism for the enal epoxidation.

CONCLUSIONS

Selected recent developments in the area of asymmetric organocatalysis in our laboratory have been briefly summarized. Enamine catalysis, Brønsted acid catalysis, and iminium catalysis turn out to be powerful new strategies for organic synthesis. Using Hantzsch ester as the hydride source, highly enantioselective transfer hydrogenantion reactions have been developed. We have also developed an additional new concept in asymmetric catalysis namely *Asymmetric Counteranion-Directed Catalysis* (ACDC) and successfully applied it to asymmetric iminium catalysis. Asymmetric induction presumably occurs in the cationic iminium ion transition state of the reaction by virtue of a sterochemical communication with the chiral phosphate counteranion, possibly via hydrogen bonding interaction. Our discovery may be of general applicability to other reactions that proceed via cationic intermediates. Despite its long roots, asymmetric organocatalysis is a relatively new and explosively growing field that, without doubt, will continue to yield amazing results for some time to come.

ACKNOWLEDGEMENT (FROM BL)

The present and past co-workers in my laboratory, whose names are given in the list of references, are highly acknowledged for their hard work, skill and enthusiasm. I thank the National Institute of Health for funding my work at Scripps. Generous support by the Max-Planck-Society, by Novartis (Young Investigator Award to BL), AstraZeneca (Award in Organic Chemistry to BL), and the Fonds der Chemischen Industrie (Kekulé fellowship to C.M.R. and Award to B.L.) is gratefully acknowledged. I also thank the DFG (Priority Program *Organocatalysis* SPP1179), Degussa, Wacker, Merck, Saltigo, Sanofi-Aventis and BASF for general support and donating chemicals.

REFERENCES

[1]   Nicolaou, K.C., Sorensen, E.J. (1996) *Classics in Total Synthesis.* Wiley-VCH, Weinheim, p. 344.

[2]   Berkessel, A., Gröger, H. (2005) *Asymmetric Organocatalysis.* Wiley-VCH, Weinheim; List, B., Yang, J.W. (2006) *Science* **313**(5793):1584.

[3]   For a review, see: Seayad, J., List, B. (2005) *Org. Biomol. Chem.* **3**:719.

[4]   a) Eder, U., Sauer, G., Wiechert, R. (1971) *Angew. Chem. Int. Ed.* **10**:496; b) Hajos, Z.G., Parrish, D.R. (1974) *J. Org. Chem.* **39**:1615. For a review, see: c) List, B. (2002) *Tetrahedron* **58**:5572.

[5]   a) Barbas, C.F. III, Heine, A., Zhong, G., Hoffmann, T., Gramatikova, S., Bjoern-
        stedt, R., List, B., Anderson, J., Stura, E.A., Wilson, I.. Lerner, R.A. (1997) *Science*
        278:2085; b) Yamada, Y.M.A., Yoshikawa, N., Sasai, H., Shibasaki, M. (1997)
        Angew. Chem. Int. Ed. **36**:1871.

[6]   a) List, B., Lerner, R.A., Barbas, C.F. III (2000) *J. Am. Chem. Soc.* **122**:2395. Also
        see: b) Notz, W., List, B. (2000) *J. Am. Chem. Soc.* **122**:7386.

[7]   a) Northrup, A.B., MacMillan, D.W.C. (2002) *J. Am. Chem. Soc.* **124**:6798;
        b) Bøgevig, A., Kumaragurubaran, N., Jørgensen, K.A. (2002) *Chem. Commun.*
        620; c) Chowdari, N.S., Ramachary, D.B., Córdova, A., Barbas, C.F. III (2002)
        Tetrahedron Lett. **43**:9591; d) Córdova, A., Notz, W., Barbas, C.F. III (2002)
        J. Org. Chem. **67**:301; e) Córdova, A., Notz, W., Barbas, C.F. III (2002) *Chem.
        Commun.* 3024; f) Sekiguchi, Y., Sasaoka, A., Shimomoto, A., Fujioka, S., Kotsuki,
        H. (2003) *Synlett.* 1655; g) Bøgevig, A., Poulsen, T.B., Zhuang, W., Jørgensen, K.A.
        (2003) *Synlett.* 1915.

[8]   a) List, B. (2000) *J. Am. Chem. Soc.* **122**:9336; b) List, B., Pojarliev, P., Biller, W.T.,
        Martin, H.J. (2002) *J. Am. Chem. Soc.* **124**:827; c) Hayashi, Y., Tsuboi, W.,
        Shoji, M., Suzuki, N. (2003) *J. Am. Chem. Soc.* **125**:11208; d) Córdova, A., Wata-
        nabe, S., Tanaka, F., Notz, W., Barbas, C.F. III (2002) *J. Am. Chem. Soc.* **124**:1866;
        e) Hayashi, Y., Tsuboi, W., Ashimine, I., Urushima, T., Shoji, M., Sakai, K. (2003)
        Angew. Chem. Int. Ed. **42**:3677; f) Córdova, A. (2003) *Synlett.* 1651; g) Notz, W.,
        Tanaka, F., Watanabe, S., Chaudari, N.S., Turner, J.M., Thayumanavan, R.,
        Barbas, C.F. III (2003) *J. Org. Chem.* **68**:9624; h) Enders, D., Grondal, C.,
        Vrettou, M., Raabe, G. (2005) *Angew. Chem. Int. Ed.* **44**:4079; i) Ibrahem, I.,
        Casas, J., Córdova, A. (2004) *Angew. Chem. Int. Ed.* **43**:6528;

[9]   a) List, B., Pojarliev, P., Martin, H.J. (2001) *Org. Lett.* **3**:2423; (b) Enders, D., Seki, A.
        (2002) *Synlett.* 26.

[10]   a) List, B. (2004) *Acc. Chem. Res.* **37**:548; b) Allemann, C., Gordillo, R.,
        Clemente, F.R., Cheong, P.H.-Y., Houk, K.N. (2004) *Acc. Chem. Res.* **37**:558.

[11]   a) List, B. (2002) *J. Am. Chem. Soc.* **124**:5656; b) Kumaragurubaran, N., Juhl, K.,
        Zhuang, W., Bøgevig, A., Jørgensen, K.A. (2002) *J. Am. Chem. Soc.* **124**:6254;
        c) Bøgevig, A., Juhl, K., Kumaragurubaran, N., Zhuang, W., Jørgensen, K.A.
        (2002) *Angew. Chem. Int. Ed.* **41**:1790.

[12]   a) Brown, S.P., Brochu, M.P., Sinz, C.J., MacMillan, D.W.C. (2003) *J. Am. Chem.
        Soc.* **125**:10808; b) Zhongm, G. (2003) *Angew. Chem. Int. Ed.* **42**:4247;
        c) Hayashi, Y., Yamaguchi, J., Hibino, K., Shoji, M. (2003) *Tetrahedron Lett.*
        44:8293; d) Hayashi, Y., Yamaguchi, J., Sumiya, T., Shoji, M. (2004) *Angew. Chem.
        Int. Ed.* **43**:1112; e) Bøgevig, A., Sundéen, H., Córdova, A. (2004) *Angew. Chem. Int.
        Ed.* **43**:1109; f) Yamamoto, H., Momiyama, N. (2005) *Chem. Commun.* 3514.

[13]   Vignola, N., List, B. (2004) *J. Am. Chem. Soc.* **126**:450.

[14]   a) Brochu, M.P., Brown, S.P., Macmillan, D.W.C. (2004) *J. Am. Chem. Soc.* **126**:4108; b) Halland, N., Braunton, A., Bachmann, S., Marigo, M., Jørgensen, K.A. (2004) *J. Am. Chem. Soc.* **126**:4790.

[15]   Enders, D., Hüttl, M.R.M. (2005) *Synlett.* 99; b) Marigo, M., Fielenbach, D., Braunton, A., Kjærsgaard, A., Jørgensen, K.A. (2005) *Angew. Chem. Int. Ed.* **44**:3703; c) Steiner, D.D., Mase, N., Barbas, C.F. III (2005) *Angew. Chem. Int. Ed.* **44**:3706; d) Beeson, T.D., MacMillan, D.W.C. (2005) *J. Am. Chem. Soc.* **127**:8826.

[16]   Bertelsen, S., Halland, N, Bachmann, S., Marigo, M., Braunton, A., Jørgensen, K.A. (2005) *Chem. Commun.* 4821.

[17]   Marigo, M., Wabnitz, T.C., Fielenbach, D., Jørgensen, K.A. (2005) *Angew. Chem. Int. Ed.* **44**:794.

[18]   Hechavarria Fonseca, M.T., List, B. (2004) *Angew. Chem. Int. Ed.* **43**:3958.

[19]   a) Pidathala, C., Hoang, L., Vignola, N., List, B. (2003) *Angew. Chem. Int. Ed.* **42**:2785; b) Tokuda, O., Kano, T., Gao, W.-G., Ikemoto, T., Maruoka, K. (2005) *Org. Lett.* **7**:5103.

[20]   a) Agami, C., Meynier, F., Puchot, C., Guilhem, J., Pascard, C. (1984) *Tetrahedron* **40**:1031; b) Agami, C., Puchot, C., Sevestre, H. (1986) *Tetrahedron Lett.* **27**:1501; c) Puchot, C., Samuel, O., Dunach, E., Zhao, S., Agami, C., Kagan, H.B. (1986) *J. Am. Chem. Soc.* **108**:2353; d) Agami, C., Puchot, C. (1986) *J. Mol. Cat.* **38**:341; e) Agami, C. (1987) *Bull. Soc. Chim. Fr.* **3**:499.

[21]   a) Hoang, L., Bahmanyar, S., Houk, K.N., List, B. (2003) *J. Am. Chem. Soc.* **125**:16; b) Klussmann, M., Iwamura, H., Mathew, S.P., Wells, D.H. Jr., Pandya, U., Armstrong, A., Blackmond, D.G. (2006) *Nature* **441**(7093*)*:621; For an more recent mechanistic study, see: c) Seebach, D., Beck, A.K., Badine, D.M., Limbach, M., Eschenmoser, A., Treasurywala, A.M., Hobi, R., Prikoszovich, W., Linder, B. (2007) *Helv. Chim. Acta* **90**:425.

[22]   a) List, B., Hoang, L., Martin, J.J. (2004) *Proc. Natl. Acad. Sci. U.S.A.* **101**:5839; b) Bahmanyar, S., Houk, K.N. (2001) *J. Am. Chem. Soc.* **123**:12911; c) Bahmanyar, S., Houk, K.N. (2001) *J. Am. Chem. Soc.* **123**:11273; d) Clemente, F.R., Houk, K.N. (2004) *Angew. Chem. Int. Ed.* **43**:5766; e) Cheong, P. H-Y., Houk, K.N. (2004) *J. Am. Chem. Soc.* **126**:13912; f) Allemann, C., Gordillom, R., Clemente, F.R., Cheong, P. H.-Y., Houk, K.N. (2004) *Acc. Chem. Res.* **37**:558; g) Bahmanyar, S., Houk, K.N., Martin, H.J., List, B. (2003) *J. Am. Chem. Soc.* **125**:2475.

[23]   a) List, B. (2001) *Synlett.* 1675; For a comprehensive review, see: b) Mukherjee, S., Yang, J.W., Hoffmann, S., List, B. (2007) *Chem. Rev.* **107**:5471 – 5569.

[24]   a) List, B., Pojarliev, P., Martin, H.J. (2001) *Org. Lett.* **3**:2423; b) Halland, N., Hazell, R.G., Jørgensen, K.A. (2002) *J. Org. Chem.* **67**:8331; c) Peelen, T.J., Chi, Y., Gellman, S.H. (2005) *J. Am. Chem. Soc.* **127**:11598; d) Chi, Y., Gellman, S.H. (2005) *Org. Lett.* **7**:4253; d) Wang, W., Wang, J., Li, H. (2005) *Angew. Chem. Int. Ed.* **44**:1369; e) Betancort, J.M., Barbas, C.F. III (2001) *Org. Lett.* **3**:3737; e) Alexakis, A., Andrey, O. (2002) *Org. Lett.* **4**:3611.

[25]   a) Yang, J.W., Stadler, M., List, B. (2007) *Angew. Chem. Int. Ed.* **46**:609; b) Yang, J.W., Stadler, M., List, B. (2007) *Nat. Protocols* **2**:1937. During our studies Enders *et al.* also reported two examples of proline-catalyzed Mannich reaction between ketones and BOC-imines: c) Enders, D., Vrettou, M. (2006) *Synthesis* 2155; d) Enders; D., Grondal, C., Vrettou, M. (2006) *Synthesis* 3597.

[26]   a) Akiyama, T., Itoh, J., Yokota, K., Fuchibe, K. (2004) *Angew. Chem. Int. Ed.* **43**:1566; b) Akiyama, T., Morita, H., Itoh, J., Fuchibe, K. (2005) *Org. Lett.* **7**:2583; also see: c) Akiyama, T. (2004) PCT Int. Appl. WO 200409675; d) Akiyama, T., Saitoh, Y., Morita, H., Fuchibe, K. (2005) *Adv. Synth. Catal.* **347**:1523–1526.

[27]   a) Uraguchi, D., Terada, M. (2004) *J. Am. Chem. Soc.* **126**:5356; b) Uraguchi, D., Sorimachi, K., Terada, M. (2004) *J. Am. Chem. Soc.* **126**:11804; c) Uraguchi, D., Sorimachi, K., Terada, M. (2005) *J. Am. Chem. Soc.* **127**:9360; also see: d) Terada, M., Uraguchi, D., Sorimachi, K., Shimizu, H. (2005) PCT Int. Appl. WO 2005070875.

[28]   a) Pictet, A., Spengler, T. (1911) *Ber.* **44**:2030; b) Tatsui, G (1928) *J. Pharm. Soc. Jpn.* **48**:92.

[29]   Taylor, M.S., Jacobsen, E.N. (2004) *J. Am. Chem. Soc.* **126**:10558.

[30]   Hagen, T.G., Narayanan, K., Names, J., Cook, J.M. (1989) *J. Org. Chem.* **54**:2170.

[31]   Seayad, J., Seayad, A.M., List, B. (2006) *J. Am. Chem. Soc.* **128**:1086.

[32]   a) Noyori, R. (2002), *Angew. Chem. Int. Ed.* **41**:2008; b) Knowles, W.S. (2002) *Angew. Chem. Int. Ed.* **41**:1998.

[33]   For reviews, see: a) Taratov, V.I., Börner, A. (2005) *Synlett* 203; b) Ohkuma, T., Kitamura, M., Noyori, R. (2000) In *Catalytic Asymmetric Synthesis*, 2nd ed.; Ojima, I., Ed.; Wiley-VCH: New York; Chapter 1; c) Ohkuma, T., Noyori, R. (2004) In *Comprehensive Asymmetric Catalysis, Suppl. 1*; Jacobsen, E.N., Pfaltz, A., Yamamoto, H., Eds.; Springer: New York, p 43. d) Nishiyama, H., Itoh, K. (2000) In *Catalytic Asymmetric Synthesis*, 2nd ed.; Ojima, I., Ed.; Wiley-VCH: New York, Chapter 2.

[34]   a) Sanwal, B.D., Zink, M.W. (1961) *Arch. Biochem. Biophys.* **94**:430; b) Kula, M.-R., Wandrey, C. (1988) *Meth. Enzymol.* **136**:34.

[35]   Itoh, T., Nagata, K., Miyazaki, M., Ishikawa, H., Kurihara, A., Ohsawa, A. (2004) *Tetrahedron* **60**:6649.

[36]   Hoffmann, S., Seayad, A.M., List, B. (2005) *Angew. Chem. Int. Ed.* **44**:7424.

[37]   Rueping, M., Sugiono, E., Azap, C., Theissmann, T., Bolte, M. (2005) *Org. Lett.* **7**:3781.

[38]   Storer, R.I., Carrera, D.E., Ni, Y., MacMillan, D.W.C. (2006) *J. Am. Chem. Soc.* **128**:84.

[39]   For reviews, see: a) Noyori, R., Tokunaga, M., Kitamura, M. (1995) *Bull. Chem. Soc. Jpn.* **68**:36; b) Ward, R.S. (1995) *Tetrahedron: Asymmetry* **6**:1475; c) Caddick, S., Jenkins, K. (1996) *Chem. Soc. Rev.* **25**:447; d) Stecher, H., Faber, K. (1997) *Synthesis* 1; e) Huerta, F.F., Minidis, A.B.E., Bäckvall, J.E. (2001) *Chem. Soc. Rev.* **30**:321; f) Perllissier, H. (2003) *Tetrahedron* **59**:8291.

[40]   Hoffmann, S., Nicoletti, M., List, B. (2006) *J. Am. Chem. Soc.* **128**:13074.

[41]   a) Li, X., List, B. (2007) *Chem. Commun.* 1739. For an independent study, see: b) Xie, J.-H., Zhou, Z.-T., Kong, W.-L., Zhou, Q.-L. (2007) *J. Am. Chem. Soc.* **129**:1868.

[42]   a) Ahrendt, K.A., Borths, C.J., MacMillan, D.W.C. (2000) *J. Am. Chem. Soc.* **122**:4243. Also see: b) Wilson, R.M., Jen, W.S., MacMillan, D.W.C. (2005) *J. Am. Chem. Soc.* **127**:11616; c) Northrup, A.B., MacMillan, D.W.C. (2002) *J. Am. Chem. Soc.* **124**:2458.

[43]   Jen, W.S., Wiener, J.J. M., MacMillan, D.W.C. (2000) *J. Am. Chem. Soc.* **122**:9874.

[44]   a) Paras, N.A., MacMillan, D.W.C. (2001) *J. Am. Chem. Soc.* **123**:4370; b) Austin, J.F., MacMillan, D.W.C. (2002) *J. Am. Chem. Soc.* **124**:1172; c) Paras, N.A., MacMillan, D.W.C. (2002) *J. Am. Chem. Soc.* **124**:7894; d) Brown, S.P., Goodwin, N.C., MacMillan, D.W.C. (2003) *J. Am. Chem. Soc.* **125**:1192.

[45]   Marigo, M., Franzen, J., Poulsen, T.B., Zhuang, W., Jørgensen, K.A. (2005) *J. Am. Chem. Soc.* **127**:6964.

[46]   Kunz, R.K., MacMillan, D.W.C. (2005) *J. Am. Chem. Soc.* **127**:3240.

[47]   a) Yang, J.W., Hechavarria Fonseca, M.T., List, B. (2004) *Angew. Chem. Int. Ed.* **43**:6660; b) Yang, J.W., Hechavarria Fonseca, M.T., Vignola, N., List, B. (2005) *Angew. Chem. Int. Ed.* **44**:108. For an independent study on the same reaction: c) Ouellet, S.G., Tuttle, J.B., MacMillan, D.W.C. (2005) *J. Am. Chem. Soc.* **127**:32.

[48]   For several reviews, see: Special edition on Asymmetric Organocatalysis (Eds. Houk, K.N. & List, B.) *Acc. Chem. Res.* 2004, **37**:487.

[49]   For a review, see: a) Lacour, J., Hebbe-Viton, V. (2003) *Chem. Soc. Rev.* **32**:373; see also: b) Llewellyn, D.B., Arndtsen, B.A. (2005) *Tetrahedron: Asymmetry* **16**:1789; c) Dorta, R., Shimon, L., Milstein, D. (2004) *J. Organomet. Chem.* **689**:751; d) Carter, C., Fletcher, S., Nelson, A. (2003) *Tetrahedron: Asymmetry* **14**:1995.

[50]   Mayer, S., List, B. (2006) *Angew. Chem. Int. Ed.* **45**:4193.

[51]   a) "Optical active citronellal" (Rhone-Poulenc Industries S.A., Fr.), JP 78–80630, 1979; b) Dang, T.-P., Aviron-Violet, P., Colleuille, Y., Varagnat, J. (1982) *J. Mol. Catal.* **16**:51; c) Kortvelyessym, G. (1985) *Acta Chim. Hung.* **119**:347.

[52]   For pioneering use of primary amine salts in asymmetric iminium catalysis involving aldehyde substrates, see: a) Ishihara, K., Nakano, K. (2005) *J. Am. Chem. Soc.* **127**:10504; b) Sakakura, A., Suzuki, K., Nakano, K., Ishihara, K. (2006) *Org. Lett.* **8**:2229. For the use of preformed imines of α,β-unsaturated aldehydes and amino acid esters in diastereoselective Michael additions, see: Hashimot, S., Komeshima, N., Yamada, S., Koga, K. (1977) *Tetrahedron Lett.* **33**:2907.

[53]   Martin, N.J.A., List, B. (2006) *J. Am. Chem. Soc.* **128**:13368.

[54]   Tuttle, J.B., Ouellet, S.G., MacMillan, D.W.C. (2006) *J. Am. Chem. Soc.* 128:12662.

[55]   Wang, X., List, B. (2008) *Angew. Chem. Int. Ed.* **47**:1119.

Systems Chemistry, May 26th – 30th, 2008, Bozen, Italy

New Tools for Molecule Makers: Emerging Technologies

Steven V. Ley[*] and Ian R. Baxendale

Department of Chemistry, University of Cambridge,
Lensfield Road, Cambridge CB2 1EW, UK

E-Mail: *svl1000@cam.ac.uk

Received: 1st October 2008 / Published: 16th March 2009

Introduction

If one reflects for a moment about the current practices used by a skilled synthesis chemist, we must be impressed by the sheer complexity of what can be achieved. Moreover, the impact on society is staggering, given the array of healing drugs, compounds that protect and guarantee our food supply to the colours and materials of our modern society. All the sciences benefit to some degree from our ability to assemble novel molecular architectures that display function and beneficial properties. The synthesis chemist's ability to understand and create these selective features at a molecular level from simple building blocks is truly awe-inspiring; especially given that a combination of only a small selection of nine different elements of the periodic table and a molecular weight limit of 500 Daltons can, in principle, generate a difficult to comprehend number of 10^{63} different molecules! Despite the obvious achievements of chemical synthesis, it is not without its problems. These relate to its current sustainability as a discipline, where we see issues of poor atom and step economy.

The rising costs, waste production and excessive use of solvents are equally unacceptable. However, over recent years many new tools have become available to molecule makers to aid them in their task [1 – 4]. These maybe biochemical in nature, such as gene shuffling techniques whereby proteins are expressed to assemble complex natural products or we could use a multitude of enzymes *via* directed evolution methods. The whole area of synthetic biology is also poised to have a significant impact. There are also synthesis techniques becoming available that allow us to work at lower scales such as the use of microarrays, mini-reactor wells and microfluidics, which we will discuss later. In addition to these, there is a range of computational methods and databases along with ReactArray reaction optimising software for design of experiments (DOE) and principal component analysis (PCA) [5 – 7]. Chemical tools for parallel synthesis or fast serial processing such as microwave methods have also become popular [8 – 12]. In our work we have concentrated on the use of immobilised reagents and scavenging methods for multi-step molecular synthesis and shown how powerful these concepts can be in the construction of pharmaceutical agents [13, 14] and natural products [15 – 20]. Indeed, it is this holistic systems chemistry approach that differentiates this from more conventional synthesis planning. In this lecture we discuss how these immobilisation methods for reagents, scavengers and catch-and-release techniques [21, 22] can be combined with phase switching and controlled release techniques to achieve chemical synthesis by continuous processing in the flow mode [23 – 25]. This will necessitate the development of suitable microreactors, packed flow tubes, flow coils, microfluidic reactor chips and appropriate reaction engineering. In order to maintain flexibility and reconfiguration of the equipment, modular units will be preferred. User-friendly interfaces and ease of operation are also important components, and although we recognise that this will represent a change in technology, it will constitute a massive change in synthesis philosophy.

General Aspects of Flow Based Chemistry

In our work, we will focus on the *quality* of the synthesis product with the aim of progressing to a multi-step operation without intermediate product isolation. While the pumping devices and reactors are clearly important, they purely assist the synthesis, and it is the processes and the quality of the product outcome that will justify their use. There is no point in simply replacing the versatile batch mode round-bottomed flasks with new tools unless clear productivity gains can be seen. Indeed, with this in mind, we can quickly list some of the anticipated synthesis opportunities and advantages of moving to flow mode. We can expect enhanced reactivity in the microfluidic channels, but also because we are working in a contained environment, high pressures and temperatures can be readily achieved. Likewise, the use of toxic, odorous or hazardous reagents would be less problematic, as would the incorporation of gases, enzymes and novel catalysts as these would be readily accommodated by the new concepts. Exothermic processes and reactive intermediates are also both readily assimilated by the devices. We can anticipate lower solvent usage and less waste-product generation leading to overall safer working practices. The machinery and processes are clearly automatable and can be easily adapted for rapid optimisation. The idea of on-demand synthesis being able to make material in the quantity and quality required at the time it is needed is also an attractive concept. Moreover, the equipment is capable of working over a 24/7 time framework, thereby extending the use and efficiency of expensive real estate. Into the future, one can imagine many further advances by using segmented flow processing for reaction and reagent profiling. Also, the idea that the machine itself, through some smart information software *via in silico* avatars, may be able to evaluate opportunities through interrogation of available literature and databases, a concept which we believe is entirely realistic. Eventually, we can anticipate linkage of several of the flow reactors, controlled by computers, to generate long synthesis sequences whereby all the modern drugs and even complex natural products will come within reach of the technology.

MICROFLUIDIC REACTORS USED IN FLOW CHEMISTRY

To begin the process of achieving flow chemistry we first describe the consequences of moving down in scale to microfluidic reactor chips which consist of serpent like channels fabricated in glass to provide the robustness required for repeated use and reuse, and to be inert to the chemistry and solvents planned in any synthesis scheme. These chips can be heated, cooled or irradiated depending on the reaction needs. Various chemical inputs can be introduced via connection ports and the reaction products can be monitored by infared or LCMS techniques in the usual way (Scheme 1).

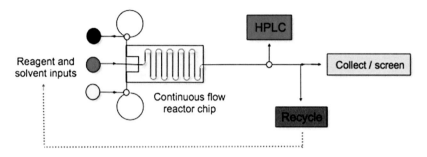

Scheme 1. Typical Flow Reactor Configuration

Recycle pathways are also possible to vary or optimise the reaction and product collection *via* preparative mass directed LC can be easily incorporated into the system. Also, importantly, small scavenger or clean-up cartridges can be linked in-line to the output stream to aid the product purity profile. In this configuration, especially with very fine channels, reactions can be very fast (seconds) when compared to similar batch mode processes (hours). When scale-up is required, it is useful to *scale-out* using multiple chips to achieve the output volumes required. A particularly important aspect of these microfluidic arrangements is that the exiting flow stream can be rapidly evaluated and the data collected can be used by intelligent feedback algorithms to iterate and improve the compound design (the closed loop concept). Especially attractive here is the ability to collect biological screening data using similar microfluidic devices immediately the compounds have been synthesised (make and screen concept). This ability to collect and assimilate data rapidly also creates wonderful synthetic opportunities for reaction, reagent or catalyst prospecting. Given that very clean products can be prepared very rapidly on this type of equipment, excellent applications to translational medicine through, for example, positron emission tomography (PET) present themselves.

As organic chemists, we expect these flow reactors to be capable of synthesis not only on a research scale, but hopefully onto full scale as well, thereby providing better continuity in the synthesis process. The equipment needs to be able to deliver pure material and avoid, where possible, high skill techniques such as chromatography, crystallisation or distillation, which are common to batch processing. In order to design systems that would fit these criteria, we devised the use of pre-packed reagent (or scavenging) tubes. These are commercially available, through Omnifit, and come in a variety of sizes, including adjustable end units, which can add further flexibility to the systems. The internal packing consists of commercially available reagents and scavengers generally supplied in bead formats. We have, however, become attracted to the use of monolithic materials prepared *in situ* as these give noticeably improved properties and loadings, and are discussed later in this article. The flow tube arrangement (Scheme 2) is versatile and can be connected and pumped in series or in parallel. Furthermore, it is possible to physically interact with the reagent through heating, cooling, oscillation, microwave, sonication and irradiation.

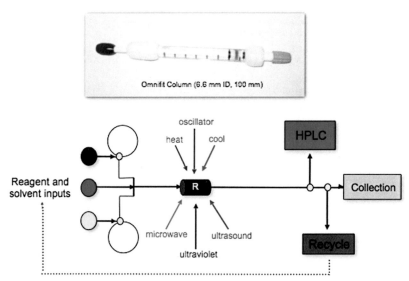

Scheme 2. Typical Flow Reactor Configuration

Recycling, reagent regeneration and reverse pumping techniques are readily accommodated. Real-time in-line analysis allows for appropriate checkpoints and valve switching so as to address complex reaction schemes, catalyst or reaction screening protocols. By way of example Scheme 3 illustrates a typical combination of chemical inputs coming together *via* an assembly of packed tubes, flow coils, chips and immobilised reagents, scavengers and catch-and-release techniques to affect the desired coupling reactions, and deal with any impurities and by-products on route to the fully constructed target molecule.

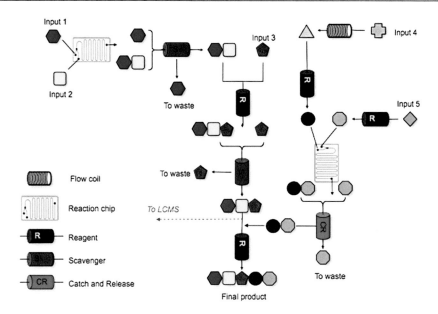

Scheme 3. Typical Flow Reactor Configuration – More Advanced Systems

FLOW SYNTHESIS EQUIPMENT

At this point it is pertinent to comment in more detail on the peripheral equipment needed to conduct these flow chemistry sequences. In its simplest form, syringe or cheap HPLC pumps can be used to drive the systems. We chose, however, to use these in combination with commercially available units from a variety of sources (see http://leyitc.ch.cam.ac.uk/) since this gives us flexibility, ease of modular assembly and effective clean-up and analysis. Also, the equipment can be tailored for specific applications such as gas reactions, microwave, chemical library synthesis, or reaction screening and profiling. While the equipment is generally used in the continuous flow mode, it should be recognised that segmental (or plug) flow operation creates additional versatility and opportunity. Although space here does not allow us to describe all of our systems in detail we would like to comment on specific items that are generally applicable. More details can be found in our publications [26 – 31] not all of which are discussed below.

The Vapourtec R4+R2 combination (Figure 1) consists of a two-pump unit, which can be controlled either manually or by computer, delivering solvent and reagents from a storage tray through various control valves to the R4 block. This is an independently temperature controlled four channel convection heater block that can be rapidly modified to use either packed reagent or scavenger flow tubes or connected to a variable length polymer or steel flow coil adaptor. These set ups can conduct syntheses ranging from milligrams potentially through to kilograms of material. Additional scavenger tubes can be added externally in-line, should the products require further purification. Units can be linked together and computer controlled to achieve multi-step mode of operation.

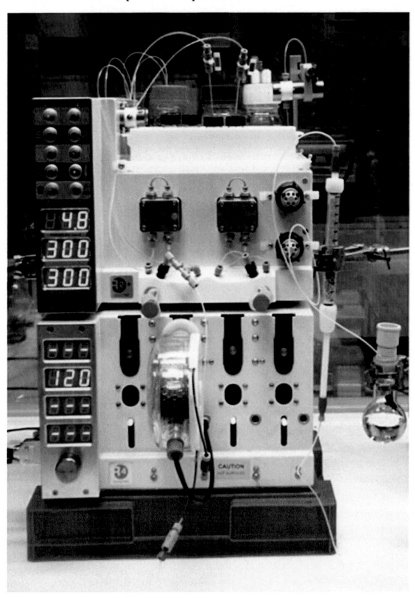

Figure 1. The Flow Coil Reactor System

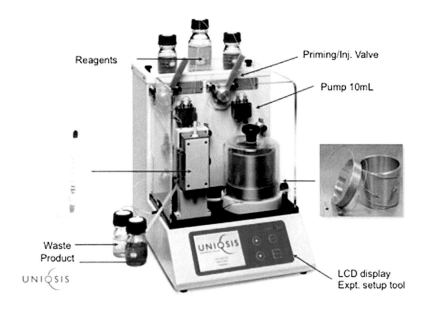

Figure 2. The Uniqsis Reactor Unit

Figure 3. Flow Synthesis Equipment Configurations

The Uniqsis FlowSyn System is an attractive alternative commercially available flow chemistry reactor (Figure 2). This employs a similar two-pump arrangement, solvent store and delivery valves with touch pad control. The primary reactor consists of a heatable steel or Peek variable length flow coil. Temperatures can be varied up to 250 °C in a covered thermal transfer heater arrangement. The exiting samples cool rapidly and can be channelled additionally through a further tube device, reaction, or scavenging as required. Alternatively, custom-built devices (Figure 3) can be assembled from a mixture of commercially available HPLC pumps, Syrris reaction chips, liquid handlers, Rheodyne valves, flow columns, detectors and product collection equipment. These flow reactors are particularly versatile and also permit intermediate sample purification through mass directed preparative chromatography, should this be deemed necessary.

APPLICATIONS OF FLOW CHEMISTRY

By way of application in this developing area of science, the Curtius reaction represents a particularly instructive example (Scheme 4). This strategically important, but hazardous to implement process converts acids through intermediate acyl azides, which rearrange at high temperature (120 °C), to useful isocyanates and their derivatives. The first sequence combines diphenylphosphoryl azide (DPPA) with a mixed stream of acid, triethylamine and a nucleophile for subsequent *in situ* trapping of the final isocyanate [30].

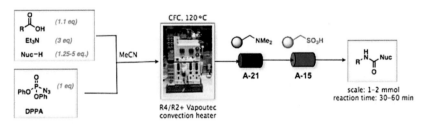

Scheme 4. Curtius Rearrangements in Flow

Passage in acetonitrile of the flow solvent through a convection-heated flow coil at 120 °C, mounted on the R4 reaction block gave a product stream that was quenched through a mixed bed of Amberlyst 21 and 15 scavengers. Alternatively, for more basic polar products, the final acidic Amberlyst 15 could be used as a "catching" device, where clean products are subsequently released by NH_3 in methanol pumped through the column. In a further modification of the Curtius process, we have shown that the intermediates acyl azides could be formed by reaction of acid chlorides flowed through an azide loaded monolith reactor [33] (Scheme 5).

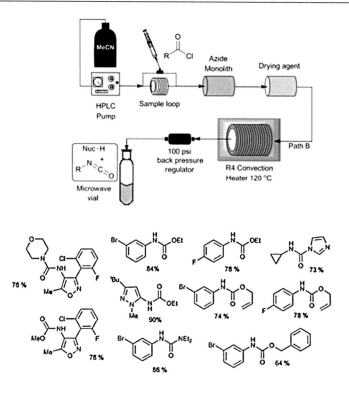

Scheme 5. Curtius Rearrangements in Flow using Azide Monoliths

In other work, owing to the thermal hazards of using azides and, to some extent, acetylenes, we have shown that the cycloaddition can be achieved smoothly by passage through a glass serpent reactor chip and onto a CuI.dimethylamino resin. Following reaction stream clean up through commercially available Quadrapure thiourea QP-TU and dibutylphosphine scavengers excellent yields and purity of the corresponding triazoles are obtained [34] (Scheme 6).

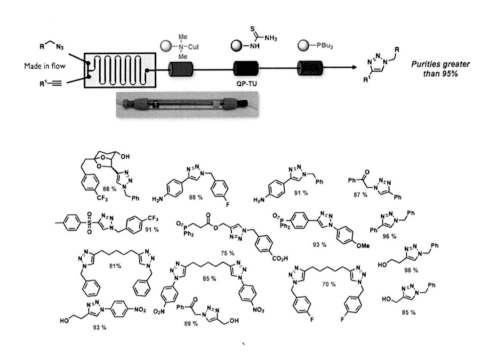

Scheme 6. Azide Couplings in Flow

By way of extending this theme of using flow reactions to control other potentially hazardous processes, we have also investigated the use of diethylaminosulfurtrifluoride (DAST) in the R4 reactor coil to achieve fluorination (Scheme 7). Several useful transformations were affected in good yield and purity, providing appropriate scavenging and product clean-up, using a plug of CaCO₃ and silica gel was incorporated into the flow sequence [35].

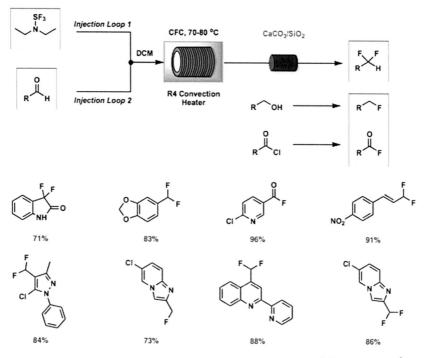

Scheme 7. Fluorination Reactions in Flow

We had mentioned earlier the improvements that accrue using monolithic polymer material to support reagents for flow chemistry. Given the importance of palladium in mediating chemical reactions, we have also shown that we can produce highly defined nanoparticles in Omnifit flow tubes [36]. These reactors are then highly efficient at achieving Heck reactions with outstanding conversion, yields and quality of product following Quadrapure scavenging (QP-TU) (Scheme 8). Clearly these metal particle reactors could be applied to a vast range of related reactions, and studies to are underway to exploit these aspects and opportunities.

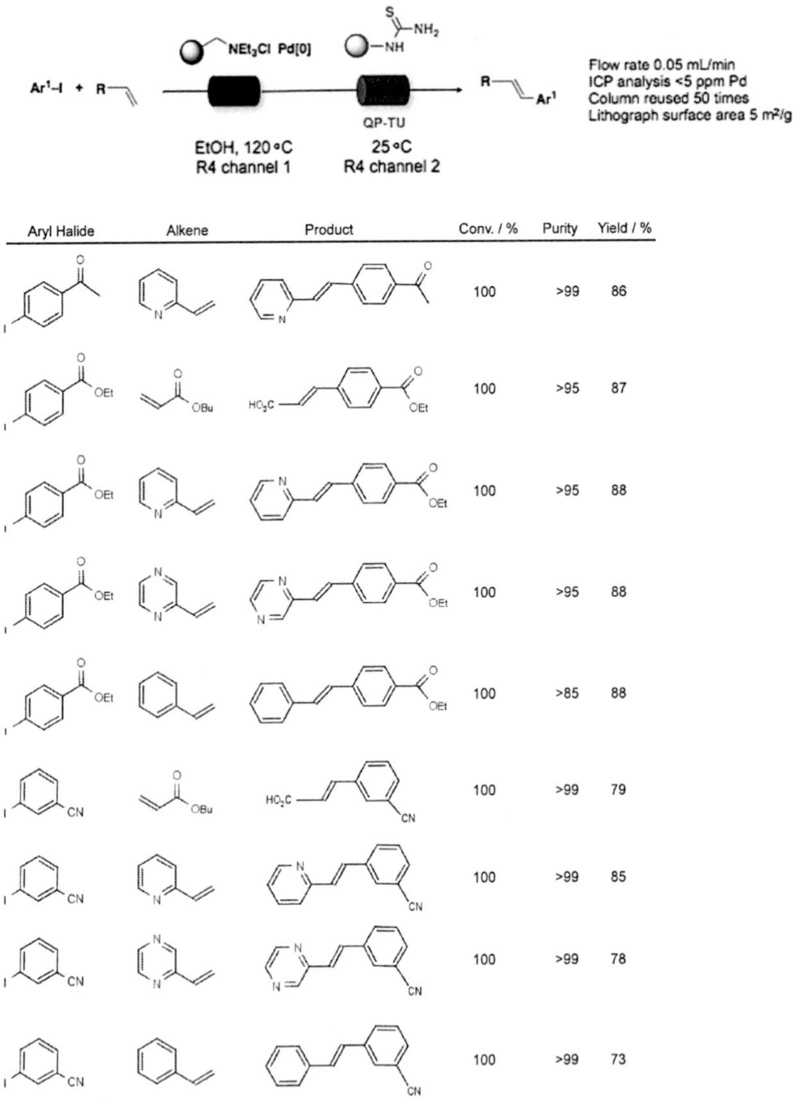

Aryl Halide	Alkene	Product	Conv. / %	Purity	Yield / %
			100	>99	86
			100	>95	87
			100	>95	88
			100	>95	88
			100	>85	88
			100	>99	79
			100	>99	85
			100	>99	78
			100	>99	73

Scheme 8. Nanoparticular Monolith Pd in Heck Reactions

NATURAL PRODUCT SYNTHESIS IN FLOW REACTORS

In other work we have reported on multi-channelled micro capillary flow tubes and shown these to be useful for several key reaction types [37]. These included Diels-Alder reactions, where we demonstrated that by scaling out to eight flow discs equivalent to 11.2 kilometres of channel, quantities of product in the order of 4 kg per day were realised. Other applications to heterocyclic synthesis [38] using flow reactor microwave methods are also possible. These methods have been effective in the generation of peptides [39] and various natural products. Two of these are reported here, since they demonstrate further interesting aspects of flow chemistry.

The first of these is the short synthesis of the neolignan natural product grossamide [40] (Scheme 9). This involved the initial coupling of an appropriately functionalised carboxylic acid, ferulic acid, by activation and trapping onto a hydroxybenzotriazole supported polymer within an Omnifit flow tube and subsequent reactions with tryptamine to give the corresponding amide. Next, after considerable experimentation with a wide range of oxidants, we found that the desired oxidative dimerisation to give grossamide in 91% yield was best accomplished using an immobilised enzyme, horseradish peroxidase, in flow mode. This last, difficult, reaction is particularly interesting in that it does not work well in batch mode due to competitive further oxidative transformations. However, by adjusting flow rates, clean chemoselective transformation occurs as the first formed grossamide product is swept through the flow tube into a benign environment where no further oxidative processes can occur. The use of immobilised enzymes in this way when combined with other chemical coupling reactions, is an attractive process for achieving molecular complexity. Considerable other opportunities arise when directed evolution techniques are used to give a novel range of enzymes for specific organic transformations. The flow equipment is ideally suited for screening and evaluation of these biotransformations.

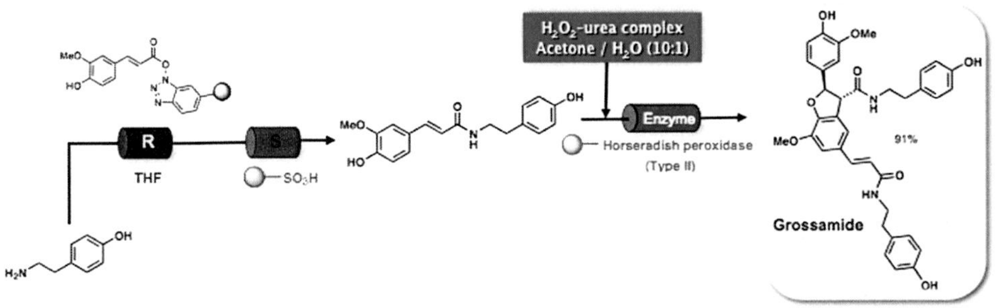

Scheme 9. Flow Synthesis of Grossamide using Immobilised Enzymes

The last synthesis describes the coupling of no less than seven chemical reactions to prepare the alkaloid natural product oxomaritidine in flow (Scheme 10) [41]. In one flow stream, a phenolic ethyl bromide was displaced by an azide delivered from an alkylammonium resin. The azide was not isolated, but simply flows to the next immobilised phosphonium reagent, where it becomes trapped as an iminophosphorane. In a second flow process, the requisite aldehyde was prepared by passage of an alcohol through an immobilised perruthenate resin [42].

The aldehyde then joins the prepared iminophosphorane, where it undergoes coupling at 55 °C to give the corresponding imine. This flow stream is then introduced to a flow hydrogenation reactor, the Thales Nano H-Cube, where reduction takes place to give an amine [43]. Solvent was then removed using a Vapourtec V10 evaporator and replaced with methylene chloride to continue the flow synthesis. Next, trifluoroacetylation of the second-ary amine was achieved by mixing with trifluoroacetic anhydride and passage through a microfluidic reactor chip. Flow through a packed column containing an immobilised hyper-valent iodine reagent [44] affects oxidative spirocyclization to a spirodienone. Finally, the synthesis was completed by deprotection of the trifluoroacetate, which causes spontaneous intramolecular conjugate addition of the amine to the enone to deliver the natural product. What is remarkable about this synthesis is that the whole process takes only a few hours to complete, which should be contrasted with the conventional batch mode preparation, which takes some 4 days of synthesis time! Clearly, the important message from this work is that the whole process can be automated to produce quality material competitively with more conventional methods, and as a result releases skilled operator time for more complex tasks such as compound and route design.

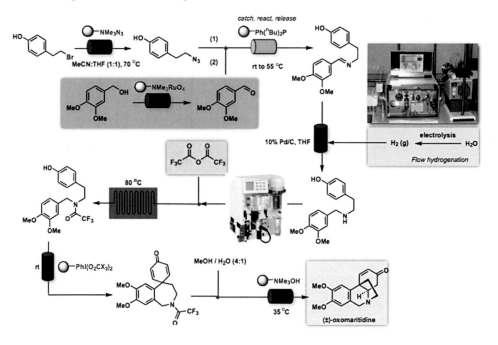

Scheme 10. Convergent Flow Synthesis of Oxomaritidine

CONCLUSION

We conclude this article by suggesting that the use of flow chemistry equipment, when combined with supported reagents, scavengers and phase switch techniques, provides a powerful platform and enabling technology for future automation of synthesis. We are not alone in this area and there is a growing user community as judged from the number of outstanding papers appearing in the literature [45 – 67]. This is, however, a disruptive technology and, as such, we cannot only expect further advances, but we can also expect resistance to adopt these new ideas. It takes time to learn the new procedures, and an open mind is essential in order to reap the eventual rewards. We have worked for over ten years with multi-step immobilised systems, including flow chemistry, and there is still much to learn. Nevertheless, it is the people that drive the changes, and the machines are only as good as the operators.

REFERENCES

[1]   Ley, S.V. and Baxendale, I.R. (2008) The Changing Face of Organic Synthesis. *Chimia* **62**:162 – 168.

[2]   Baxendale, I.R., Hayward, J.J. and Ley, S.V. (2007) Microwave Reactions Under Continuous Flow Conditions. *Comb. Chem. And High Throughput Screening* **35**:802 – 836.

[3]   Baxendale, I.R., Hayward, J.J., Ley, S.V. and Tranmer, G.K. (2007) Pharmaceutical Strategy and Innovation: An Academic's Perspective. *Chem. Med. Chem.* **2(6)**:768 – 788.

[4]   Baxendale, I.R. and Ley, S.V. (2002) New Tools and Concepts for Modern Organic Synthesis. *Nat. Rev. Drug Disc.* **1**:573 – 586.

[5]   Jamieson, C., Congreave, M.S., Emiabata-Smith, D.F. and Ley, S.V. (2000) A Rapid Approach for the Optimisation of Polymer-Supported Reagents in Synthesis. *Synlett.* 1603 – 1607.

[6]   Jamieson, C., Congreave, M.S., Emiabata-Smith, D.F., Ley, S.V. and Scicinski, J.J. (2002) Application of ReactArray Robotics and Design of Experiments Techniques in Optimisation of Supported Reagent Chemistry. *Organic Process. Res. and Dev.* **6**:823 – 825.

[7]   Scicinski, J.J., Congreave, M.C., Jamieson, C., Ley, S.V., Newman, E.S., Vinader, V.M. and Carr, R.A.E. (2001) Solid-Phase Development of a 1-Hydroxy-benzotriazole Linker for Heterocycle Synthesis using Analytical Constructs. *J. Combinatorial Chem.* **3**:387 – 396.

[8]   Baxendale, I.R., Lee, A.-L. and Ley, S.V. (2005) Integrating Microwave-Assisted Synthesis and Solid-Supported Reagents. In: *Microwave-Assisted Organic Synthesis*, (Tierney, J.P., Lidstrom, P., Eds.), pp.133 – 176. Blackwells.

[9]   Ley, S.V., Baxendale, I.R., Brusotti, G., Massi, A. and Nesi, M. (2002) Solid-Supported Reagents for Multi-Step Organic Synthesis: Preparation and Application. *Il Farmaco* **57**:321 – 330.

[10]   Baxendale, I.R. and Ley, S.V. (2000) Polymer-Supported Reagents for Multistep Organic Synthesis: Application to the Synthesis of Sildenafil. *Bioorg. Med. Chem. Lett.* **10**:1983 – 1986.

[11]   Baxendale, I.R., Hayward, J.J., Ley, S.V. (2007) Microwave reactions under continuous flow conditions. *Comb. Chem. High Throughput Screening* **35**:802 – 836.

[12]   Baxendale, I.R., Pitts, M.R. (2006) Microwave flow chemistry: the next evolutionary step in synthetic chemistry? *Chem. Today* **24**:41 – 45.

[13]   Ley, S.V., Baxendale, I.R., Longbottom, D.A. and Myers, R.M. Ed. Chorghade, M.S. (2007) Natural Products as an Inspiration for the Discovery of New High Throughput Chemical Synthesis Tools. In: *Drug Discovery & Development, Vol. 2 Drug Development.* (Chorghade, M.S., Ed.), pp.51 – 89. John Wiley & Sons.

[14]   Siu, J., Baxendale, I.R. and Ley, S.V. (2004) *Org. Biomol. Chem.* **2**:160 – 167.

[15]   Ley, S.V., Baxendale, I.R. and Myers, R.M. (2006) Polymer-Supported Reagents and Scavengers in Synthesis. In: *Comprehensive Medicinal Chemistry II, Vol.3, Drug Discovery Technologies.* (Triggle, D.J. and Taylor, J.B., Eds.), pp.791 – 836. Elsevier, Oxford.

[16]   Baxendale, I.R. and Ley, S.V. (2005) Synthesis of Alkaloid Natural Products Using Solid-Supported Reagents and Scavengers. *Curr. Org. Chem.* **9**:1521 – 1534.

[17]   Storer, R.I., Takemoto, T., Jackson, P.S., Brown, D.S., Baxendale, I.R. and Ley, S.V. (2004) Multi-Step Application of Immobilised Reagents and Scavengers: A Total Synthesis of Epothiolone C. *Chem. Eur. J.* **10**:2529 – 2547.

[18]   Baxendale, I.R., Davidson, T.D., Ley, S.V. and Perni, R.H. (2003) Enantioselective Synthesis of the Tetrahydrobenzylisoquinoline Alkaloid (-)-Norarmepavine Using Polymer Supported Reagents. *Heterocycles* **60**:2707 – 2715.

[19]   Baxendale, I.R., Ley, S.V. and Piutti, C. (2002) Total Synthesis of the Amaryllidaceae Alkaloid (+)-Plicamine and Its Unnatural Enantiomer by Using Solid-Supported Reagents and Scavengers in a Multistep Sequence of Reactions. *Angew. Chem. Int. Ed.* **41**:2194 – 2197.

[20]   Baxendale, I.R., Lee, A.-L. and Ley, S.V. (2002) A Concise Synthesis of Carpanone using Solid-Supported Reagents and Scavengers. *J. Chem. Soc. Perkin Trans 1.* **16**:1850 – 1857.

[21]   Ley, S.V., Baxendale, I.R., Bream, R.N., Jackson, P.S., Leach, A.G., Longbottom, D.A., Nesi, M., Scott, J.S., Storer, R.I. and Taylor, S.J. (2000) Multistep Organic Synthesis using Solid Supported Reagents and Scavengers: A New Paradigm in Chemical Library Generation. *J. Chem. Soc. Perkin. Trans.* 1:3815 – 4195.

[22]   Baxendale, I.R., Storer, R.I. and Ley, S.V. (2003) Supported Reagents and Scavengers in Multi-Step Organic Synthesis. In: *Polymeric Materials in Organic Synthesis and Catalysis.* (Buchmeiser, M.R. Ed.), pp.53 – 136. VCH Berlin.

[23]   Baxendale, I.R. and Ley, S.V. (2007) Solid Supported Reagents in Multi-Step Flow Synthesis. In: *New Avenues to Efficient Chemical Synthesis: Emerging Technologies, Symposium   Proceeding   06.3.*   (Seeberger,   P.H.   and   Blume,   T.,   Eds), pp. 151 – 185. Springer, Berlin Heidelberg.

[24]   Baxendale, I.R., Hayward, J.J., Lanners, S., Ley, S.V. and Smith, C.D. (2008) Organic Chemistry in Microreactors; Heterogenous Reactions. In: *Microreactors in Organic Chemistry and Catalysis.* (Wirth, T. and Wiley, J., Eds), pp. 84 – 122. Wiley-VCH.

[25]   Ley, S.V and Baxendale, I.R. (2008) The Changing Face of Organic Synthesis. *Chimia* **62**:162 – 168.

[26]   Smith, C.J., Iglesias-Sigüenza, F.J., Baxendale, I.R. and Ley, S.V. (2007) Flow and batch mode focused microwave synthesis of 5-amino-4-cyanopyrazoles and their further   conversion   to   4-aminopyrazolopyrimidines.   *Org.   Biomol.   Chem.* **5**:2758 – 2761.

[27]   Smith, C.D., Baxendale, I.R., Tranmer, G.K., Baumann, M., Smith, S.C., Lewthwaite, R.A. and Ley, S.V. (2007) Tagged phosphine reagents to assist reaction work-up by phase-switched scavenging using a modular flow reactor. *Org. Biomol. Chem.* **5**:1562 – 1568.

[28]   Baxendale, I.R., Ley, S.V., Smith, C.D. and Tranmer, G.K. (2006) A flow reactor process for the synthesis of peptides utilizing immobilized reagents, scavengers and catch and release protocols. *Chem. Commun.* **46**:4835 – 4837.

[29]   Baxendale, I.R., Griffiths-Jones, C.M., Ley, S.V. and Tranmer, G.K. (2006) Microwave-Assisted Suzuki Coupling Reactions with an Encapsulated Palladium Catalyst for Batch and Continuous-Flow Transformations. *Chem. Eur. J.* **12**(16):4407

[30]   Saaby, S., Baxendale, I.R. and Ley, S.V. (2005) Non-metal-catalysed intramolecular alkyne cyclotrimerization reactions promoted by focussed microwave heating in batch and flow modes. *Org. Biomol. Chem.* **3**:3365 – 3368.

[31] Ley, S.V., Ramarao, C., Gordon, R.S., Holmes, A.B., Morrison, A.J., McConvey, I.F., Shirley, I.M., Smith, S.C. and Smith, M.D. (2002) Polyurea-encapsulated palladium(II) acetate: a robust and recyclable catalyst for use in conventional and supercritical media. *Chem. Commun.* **10**:1134 – 1135.

[32] Baumann, M., Baxendale, I.R., Ley, S.V., Nikbin, N., Smith, C.D. and Tierney, J.P. (2008) A Modular Flow Reactor for Performing Curtius Rearrangements as a Continuous Flow Process. *Org. Biomol. Chem.* **6**:1577 – 1586.

[33] Baumann, M., Baxendale, I.R., Ley, S.V., Nikbin, N., Smith, C.D. (2008) Azide Monoliths as Convenient Flow Reactors for Efficient Curtius Rearrangement Reactions. *Org. Biomol. Chem.* **6**:1587 – 1593.

[34] Smith, C.D., Baxendale, I.R., Lanners, S., Hayward, J.J., Smith, S.C. and Ley, S.V. (2007) [3 + 2] Cycloaddition of Acetylenes with Azides to give 1,4-Disubstituted 1,2,3- Triazoles in a Modular Flow Reactor. *Org. Biomol. Chem.* **5**:1559 – 1561.

[35] Baumann, M., Baxendale, I.R. and Ley, S.V. (2008) The Use of Diethylaminosulfur Trifluoride (DAST) for Fluorination in a Continuous-Flow Microreactor. *Synlett.* **14**:2111 – 2114.

[36] Nikbin, N., Ladlow, M. and Ley, S.V. (2007) Continuous Flow Ligand-Free Heck Reactions Using Monolithic Pd[0] Nanoclusters. *Org. Proc. Res. Dev.* **11**:458 – 462.

[37] Hornung, C. H., Mackley, M. R., Baxendale I. R., and Ley, S.V. (2007) A Microcapillary Flow Disc (MFD) Reactor for Organic Synthesis. *Org. Proc. Res. Dev.* **11**:399 – 405.

[38] Baumann, M., Baxendale, I.R., Ley, S.V., Smith, C.D. and Tranmer, G.K. (2006) A Fully Automated Continuous Flow Synthesis of 4,5-Disbustituted Oxaxoles. *Org. Lett.* **8**:5231 – 5234.

[39] Baxendale, I.R., Ley, S.V., Smith C.D. and Tranmer, G.K. (2006) A Flow Reactor Process for the Synthesis of Peptides Utilizing Immobilised Reagents, Scavengers and Catch and Release Protocols. *J. Chem. Soc. Chem. Commun.* 4835 – 4837.

[40] Baxendale, I.R., Griffiths-Jones, C.M., Ley, S.V. and Tranmer, G.K. (2006) Preparation of the Neolignan Natural Product Grossamide by a Continuous Flow Process. *Synlett.* 427 – 430.

[41] Baxendale, I.R., Deeley, J., Griffiths-Jones, C.M., Ley, S.V., Saaby,S. and Tranmer, G. (2006) A Flow Process for the Multi-Step Synthesis of the Alkaloid Natural Product Oxomaritidine: A New Paradigm for Molecular Assembly. *J. Chem. Soc. Chem. Commun.* 2566 – 2568.

[42] Hinzen B., and Ley, S.V., (1997) Polymer-Supported Perruthenate (PSP): A New Oxidant for Clean Organic Synthesis. *J. Chem. Soc. Perkin Trans.* **1**:1907 – 1908.

[43] Saaby, S., Knudsen, K.R., Ladlow, M. and Ley, S.V. (2005) The Use of a Continuous Flow-Reactor Employing a Mixed Hydrogen-Liquid Flow Stream for the Efficient Reduction of Imines to Amines. *J. Chem. Soc. Chem. Commun.* 2901–2911

[44] Ley, S.V., Thomas, A.W., and Finch, H. (1999) Polymer-Supported Hypervalent Iodine Reagents in 'Clean' Organic Synthesis with Potential Applications in Combinatorial Chemistry. *J. Chem. Soc. Perkin Trans.* **1**:669–671.

[45] Jun-ichi Yoshida, Nagaki, A. and Yamada, T. (2008) Flash Chemistry: Fast Chemical Synthesis by Using Microreactors. *Chemistry – A European Journal* **14**:7450–7459.

[46] McConnell, J.R., Hitt, J.E., Daugs, E.D. and Rey, T.A. (2008) The Swern Oxidation: Development of a High-Temperature Semicontinuous Process. *Org. Process Res. Dev.* **12**:940–945.

[47] Acke, D.R.J., Stevens, C.V. and Roman, B.I. (2008) Microreactor Technology: Continuous Synthesis of 1 H-Isochromeno[3,4-d]imidazol-5-ones. *Org. Process Res. Dev.* **12**:921–928.

[48] Goto, S., Velder, J., Sheikh, S. E., Sakamoto, Y., Mitani, M., Elmas, S., Adler, A., Becker, A., Neudörfl, J.-M., Lex, J. and Schmalz, H.-G. (2008) Butyllithium-Mediated Coupling of Aryl Bromides with Ketones under In-Situ-Quench (ISQ) Conditions: An Efficient One-Step Protocol Applicable to Microreactor Technology. *Synlett.* **9**:1361–1365.

[49] Gustafsson, T., Gilmour, R. and Seeberger, P.H. (2008) Fluorination reactions in microreactors. *Chem. Commun.* **26**:3022–3024.

[50] Kraai, G.N., Zwol, F.V., Schuur, B., Heeres, H.J. and Vries, J.G.D. (2008) Two-Phase (Bio)Catalytic Reactions in a Table-Top Centrifugal Contact Separator. *Angew. Chem. Int. Ed.* **47**:3905–3908.

[51] Tanaka, K., Motomatsu, S., Koyama, K. and Fukase, K. (2008) Efficient aldol condensation in aqueous biphasic system under microfluidic conditions. *Tetrahedron Lett.* **49**:2010–2012.

[52] Burguete, M.I., Erythropel, H., Garcia-Verdugo, E., Luis, S.V. and Sans, V. (2008) Base supported ionic liquid-like phases as catalysts for the batch and continuous-flow Henry reaction. *Green Chem.* **10**:401–407.

[53] Csajági, C., Szatzker, G., Ťke, E.R., Ürge, L., Darvas, F.and. Poppe, L. (2008) Enantiomer selective acylation of racemic alcohols by lipases in continuous-flow bioreactors. *Tetrahedron: Asymmetry* **19**:237–246.

[54] Gustafsson, T., Pontén, F. and Seeberger, P.H. (2008) Trimethylammonium mediated amide bond formation in a continuous flow microreactor as key to the synthesis of rimonabant and efaproxiral. *Chem. Commun.* 1100–1102.

[55]   Fukuyama, T., Kobayashi, M., Rahman, M.T., Kamata, N. and Ryu, I. (2008) Spurring Radical Reactions of Organic Halides with Tin Hydride and TTMSS Using Microreactors. *Org. Lett.* **10**:533–536.

[56]   Chambers, R.D., Sandford, G., Trmcic, J. and Okazoe, T. (2008) Elemental Fluorine. Part 21. Direct Fluorination of Benzaldehyde Derivatives. *Org. Process Res. Dev.* **12**:339–344.

[57]   Ushiogi, Y., Hase, T., Iinuma, Y., Takata, A. and Yoshida, J. (2007) Synthesis of photochromic diarylethenes using a microflow system. *Chem. Commun.* **28**:2947–2949.

[58]   Lozano, P., García-Verdugo, E., Piamtongkam, R., Karbass, N., Diego, T.D., Burguete, M.I., Luis, S.V. and Iborra, J.L. (2007) Bioreactors Based on Monolith-Supported Ionic Liquid Phase for Enzyme Catalysis in Supercritical Carbon Dioxide. *Adv. Synth. Catal.* **349**:1077–1084.

[59]   Dräger, G., Kiss, C., Kunz, U. and Kirschning, A. (2007) Enzyme-purification and catalytic transformations in a microstructured PASSflow reactor using a new tyrosine-based Ni-NTA linker system attached to a polyvinylpyrrolidinone-based matrix. *Org. Biomol. Chem.* **5**:3657–3664.

[60]   Hartung, A., Keane, M.A. and Kraft, A. (2007) Advantages of Synthesizing trans-1,2-Cyclohexanediol in a Continuous Flow Microreactor over a Standard Glass Apparatus. *J. Org. Chem.* **72**:10235–10238.

[61]   Kralj, J.G., Sahoo, H.R. and Jensen, K.F. (2007) Integrated continuous microfluidic liquid–liquid extraction. *Lab Chip* **7**:256–263.

[62]   Kawanami, H., Matsushima, K., Sato, M. and Ikushima, Y. (2007) Rapid and Highly Selective Copper-Free Sonogashira Coupling in High-Pressure, High-Temperature Water in a Microfluidic System. *Angew. Chem. Int. Ed.* **46**:5129–5132.

[63]   Hamper, B.C. and Tesfu, E. (2007) Direct Uncatalyzed Amination of 2-Chloropyridine Using a Flow Reactor. *Synlett* **14**:2257–2261.

[64]   Murphy, E.R., Martinelli, J.R., Zaborenko, N., Buchwald, S.L. and Jensen, K.F. (2007) Accelerating Reactions with Microreactors at Elevated Temperatures and Pressures: Profiling Aminocarbonylation Reactions. *Angew. Chem. Int. Ed.* **46**:1734–1737.

[65]   Burguete, M.I., Cornejo, A., García-Verdugo, E., Gil, M.J., Luis, S.V., Mayoral, J.A., Martínez-Merino, V. and Sokolova, M. (2007) Pybox Monolithic Miniflow Reactors for Continuous Asymmetric Cyclopropanation Reaction under Conventional and Supercritical Conditions. *J. Org. Chem.* **72**:4344–4350.

[66]   Solodenko, W., Jas, G., Kunz, U. and Kirschning, A. (2007) Continuous Enantiose-
        lective Kinetic Resolution of Terminal Epoxides Using Immobilized Chiral Cobalt-
        Salen Complexes. *Synthesis* **4**:583 – 589.

[67]   Sahoo, H.R., Kralj, J.G. and Jensen, K.F. (2007) Multistep Continuous-Flow Micro-
        chemical Synthesis Involving Multiple Reactions and Separations. *Angew. Chem. Int.
        Ed.* **46**:5704 – 5708.

MICROREACTORS AS THE KEY TO THE CHEMISTRY LABORATORY OF THE FUTURE

KAROLIN GEYER AND PETER H. SEEBERGER [*,1]

Laboratory of Organic Chemistry,
Swiss Federal Institute of Technology (ETH) Zurich
Wolfgang-Pauli-Strasse 10, 8093 Zurich, Switzerland

E-Mail: *seeberger@org.chem.ethz.ch

Received: 4th September 2008 / Published: 16th March 2009

ABSTRACT

The aim of synthetic chemists and the chemical industry is to perform chemical transformations in a highly efficient and environmentally benign manner. This involves atom economy and atom efficiency of the particular reactions on one hand; on the other hand it includes aspects of the reaction performance such as process safety, solvent and reagent consumption, and purification procedures. A well-engineered approach to elegantly overcome some of the aspects related to process performance is the use of continuous-flow microfluidic devices in chemical laboratories. This chapter highlights some application of these new tools in synthetic laboratories and how continuous-flow reactors may change the way chemists will perform their research in the future.

INTRODUCTION

Research chemists typically perform chemical transformations in traditional glass round bottomed flasks. Since the optimization of chemical transformations in batch reactors often consumes substantial amounts of starting materials, a lot of precious building blocks as well as effort and time are required in order to identify the ideal reaction conditions for a

---

[1] New adress: Max-Planck-Institute of Colloids and Interfaces, Department of Biomolecular Systems, Am Mühlenberg 1, 14476 Potsdam-Golm, Germany; E-Mail: peter.seeberger@mpikg.mpg.de

particular reaction. Having found the optimal conditions to achieve a certain reaction on a small scale, process scale-up often poses additional challenges and requires further adjustment of the reaction parameters. To overcome these hurdles in synthetic chemistry, microstructured continuous-flow reactors and chip-based microreactors are becoming increasingly popular [1].

Microstructured continuous-flow devices consist of a miniaturized channel system etched into or established on materials such as metals, silicon, glass, ceramics or polymers, which allows the chemical reaction to take place in a relatively narrow pore. These reaction systems have a high interfacial area per volume (only depending on the radius of the channel) and chemical reactions therefore profit from rapid heat and mass transfer. Due to the high heat and mass transfer rates, the reaction time, yield and selectivity are strongly influenced, often rendering processes to be more efficient and selective and thereby avoiding the generally required purification processes. Besides the facile control of the physical parameters, microreactors allow for low operational volumes to minimize reagent consumption, the integration of *on-line* detection modules and an excellent process safety profile in case highly exothermic reactions are undertaken, enhanced by the fact that only small amounts of hazardous/explosive intermediates may be formed at any given time. To obtain synthetically useful amounts of product, the reactors are simply run longer ("scale-out" principle [1e]) or several reactors are placed in parallel ("numbering up"), assuring identical conditions for the analytical and preparative modes. Many different applications of microstructured devices in synthetic chemistry have been reported and reviewed; [2, 3] here some concepts and recent developments from our laboratory will be presented.

MICROFLUIDIC REACTOR TYPES

Microreactors consist of a network of miniaturized channels often embedded in a flat surface, referred to as the "chip" [1a]. Since different chemical applications call for different types of reactors and materials, a variety of different possibilities have been explored. Additionally, the size of microfluidic devices strongly influences their application and the way in which reagents are introduced to the system [4]. The features of several selected microreactors are summarized in Figure 1 [5].

Most commonly, glass, stainless steel, silicon or polymers (e. g. poly(dimethylsiloxane), PDMS or poly(tetrafluoroethylene), PTFE) are used. Devices fabricated of PDMS are traditionally employed in biological and biochemical applications, [11] whereas stainless steel based microreactors are mostly applied in *meso*-scale production efforts [2b, 12]. Glass devices allow for the visual inspection of the reaction progress [13] or for application to photochemical reactions, [14] whereas silicon reactors offer excellent heat transfer rates due to the high thermal conductivity of silicon and therefore often are the reactors of choice due to their well established fabrication procedures [2 d, 3b–e, 15].

Figure 1. Selected microreactors; Top row from left to right: Silicon-based micro-reactor designed by Jensen [3b, 10]; Glass microreactor by Syrris® [6]; Stainless steel microreactor system by Ehrfeld Mikrotechnik® [7]. Bottom row from left to right: Glass microreactor by Haswell [1b, 1 l]; Stainless steel microreactor of the CYTOS® Lab system[8]; Glass microreactor by Micronit Microfluidics®[9]; Large picture right: Vapourtec® tube reactor [10].

The initial investigations of our group employed a stainless steel microreactor system designed by Ehrfeld Mikrotechnik [3a, 7] (see Figure 1). We further focused on the application of silicon-glass microreactors [3b-e, 15] as the glass layer would offer the possibility to visually inspect the reaction and the silicon portion would provide a rapid heat transfer. Further investigations were carried out in commercially available glass and PTFE tube microreactors [3f-h, 6], and other system are currently under evaluation or being developed.

SYNTHESIS IN MICROCHEMICAL SYSTEMS

Due to the small channel dimensions and the increased surface to volume ratio of micro-reactors, mass and heat transport are significantly more efficient than in the classic round-bottomed flask. The mixing of reagents by diffusion occurs very quickly, and heat exchange between the reaction medium and reaction vessel is highly efficient. As a result, the reaction conditions in a continuous-flow microchannel are homogenous, and can be controlled precisely. Therefore, highly exothermic and even explosive reactions can be readily harnessed in a microreactor. The careful control of reaction temperature and residence time has a beneficial effect on the outcome of a reaction with respect to yield, purity and selectivity. Below, selected examples from our laboratory are described to illustrate the potential application of microreactors to organic synthesis. The application of microreactors to small-scale medicinal and academic total synthesis is just beginning.

CHEMICAL REACTIONS: INITIAL INVESTIGATIONS AND SYNTHESIS OF BIOPOLYMERS

Liquid-phase reactions performed in micro structured devices benefit particularly from the efficient mass and heat transport characteristics of microreactors and the small amounts of reactants (e. g. hazardous reagent or intermediates, precious starting materials) being present in the system at any given time.

Our laboratory initially explored the relevance of microreactor technology in synthetic chemistry employing a stainless steel microreactor system (see Figure 1, Scheme 1) [3a]. The system was used to undertake various important synthetic transformations such as Williamson ether synthesis (eq. 1), Diels-Alder reactions (eq. 2), Horner-Wadsworth-Emmons reactions (eq. 3) and Henry reactions (eq. 4). Since a significant number of chemical transformations require solid catalysts to proceed in a reasonable reaction time and selectivity, investigations were undertaken to perform reactions of a heterogeneous nature in continuous-flow. A pre-packed reaction cartridge, loaded with an active catalyst on polymer beads, was placed in the continuous-flow system to allow palladium mediated cross-coupling reactions (Heck reaction, eq. 5) to take place [3a].

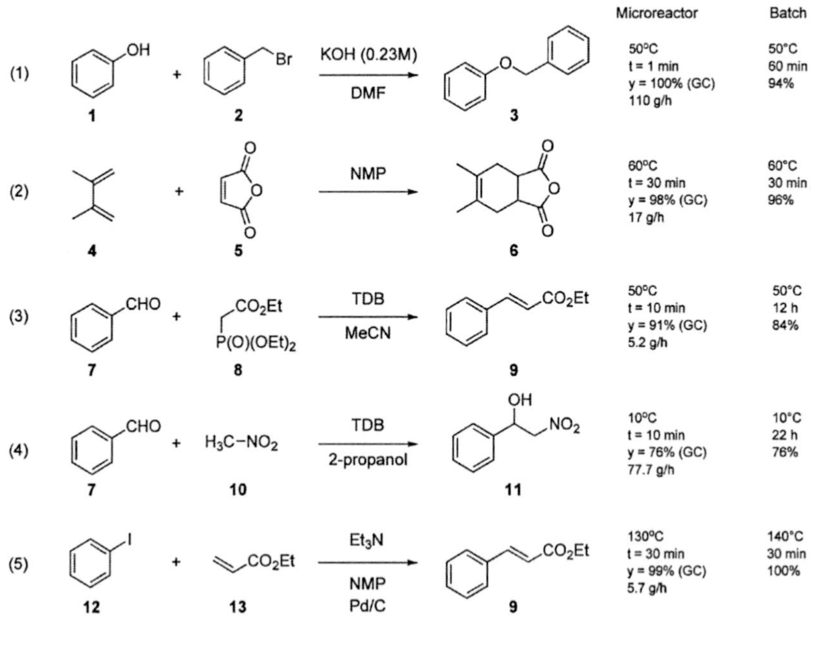

Scheme 1. Selected transformations carried out in a stainless steel microreactor system (y = yield) [3a].

The chemical synthesis of carbohydrate building blocks or small oligosaccharides remains a challenging task for synthetic chemists since a huge variety of components influence the outcome of a given glycosylation [3b, d]. The chemical outcome of such transformations depends strongly on the steric and electronic properties of the coupling partners, reagent concentrations, reaction temperature and reaction time. Furthermore, the building blocks used for oligosaccharide assembly often require multistep syntheses and are precious synthetic intermediates themselves. A silicon-glass microreactor, allowing for careful control of the reaction parameters, was employed to investigate glycosylations in continuous-flow (Scheme 2) [3b]. The small internal volume of 78 µL minimized reagent and building block consumption. The reaction progress of the coupling between protected mannoside 14 and galactoside 15 was monitored as a function of time and temperature (Fig. 2).

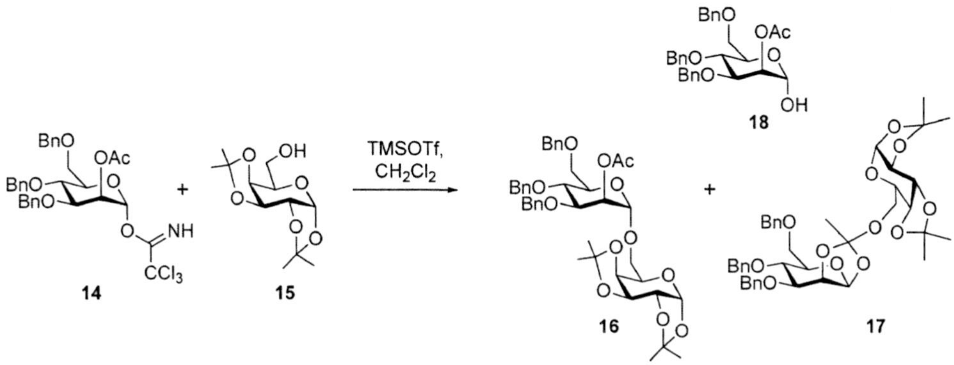

Scheme 2. Initial application of the silicon-glass microreactor to carbohydrate chemistry [3b].

It was revealed that at low temperatures ($-80\,°C$ to $-70\,°C$) and short reaction times (<1 min) the formation of orthoester 17 was favored, whereas higher temperatures ($-40\,°C$) and longer reaction times (~ 4 min) led to formation of the desired α-linked product 16. Using as little as 100 mg of starting materials, 40 different reaction conditions were scanned within one day [3b].

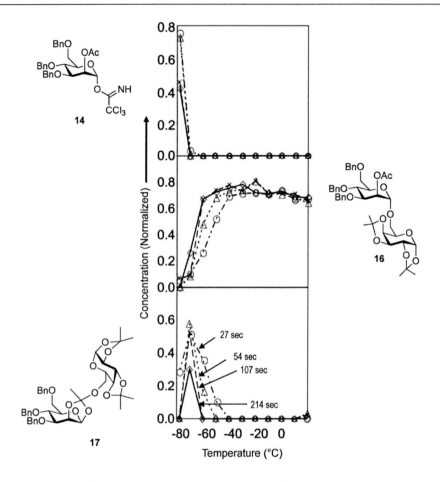

Figure 2. Starting material consumption and product formation of a glycosylation reaction [3b].

Having proven the applicability of the silicon-glass microreactor in carbohydrate chemistry, we were interested in investigating the influence of reaction temperature, reagent concentration, solvent and anomeric leaving group on the *selective* outcome of a given glycosylation. Therefore comparative studies employing mannosyl building blocks **14**, **19** and **20** respectively were undertaken to screen for the optimal reaction conditions varying reaction time, temperature, solvent and reagent concentration (Scheme 3) [3 d].

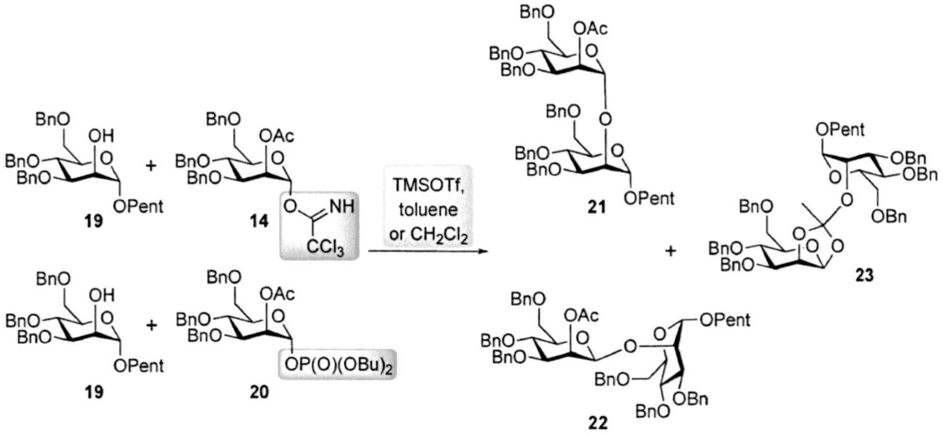

Microreactors as the Key to the Chemistry Laboratory of the Future

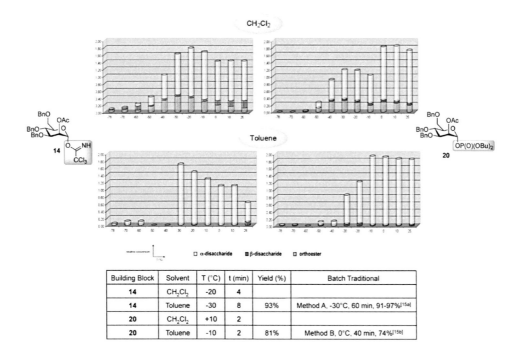

Scheme 3. Comparative studies of glycosylation reactions using varying glycosylating agents and solvent system [3 d].

Building Block	Solvent	T (°C)	t (min)	Yield (%)	Batch Traditional
14	CH₂Cl₂	-20	4		
14	Toluene	-30	8	93%	Method A, -30°C, 60 min, 91-97%[15a]
20	CH₂Cl₂	+10	2		
20	Toluene	-10	2	81%	Method B, 0°C, 40 min, 74%[15b]

Figure 3. Product distribution at the optimal reaction conditions found by screening; Left side: Glycosylations employing reactant **14**; Right side: Glycosylations employing reactant **20**; **Table:** Scale-out of the optimized procedures [3 d].

It was discovered that glycosylations in toluene as the solvent proceeded more selectively than in dichloromethane (Fig. 3). While in dichloromethane even at optimized reaction conditions undesired β-disaccharide **22** and orthoester **23** are formed, α-disaccharide **21** is exclusively formed in toluene. Scale-up of the established synthetic procedures yielded the desired saccharide **21** in good to excellent yields (81% and 93% respectively) in short reaction times (2 min, 8 min) [3 d].

After having established the reaction conditions for single glycosylation reactions in silicon-glass microreactors, the synthesis of an oligosaccharide in continuous-flow was explored [3e]. Glycosylphosphate **24** was iteratively coupled to a growing carbohydrate chain on a perfluorinated support linker for facile purification by fluorous solid phase extraction (FSPE [17]) to deliver the desired oligosaccharides (Scheme 4). *In-situ* deprotection of the primary alcohol of each carbohydrate allowed for a reaction sequence of coupling and deprotection within reaction times of up to one minute per sequence to obtain the desired oligosaccharide in excellent yield (Scheme 5). The perfluorinated support linker of homotetrasaccharide **29** could, after successful assembly of the structure, be further functionalized to yield terminal olefins, aldehydes and trichloroacetimidates allowing for further synthetic transformations and attachment onto slides for biological investigations respectively (Scheme 6) [3e].

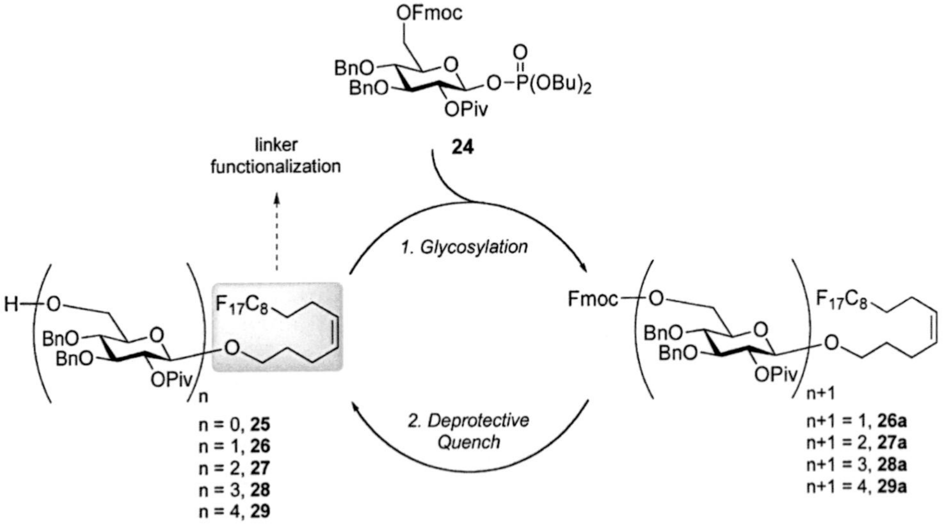

Scheme 4. Synthesis of the glucose based homotetramer **29** via iterative glycosylation [3 e].

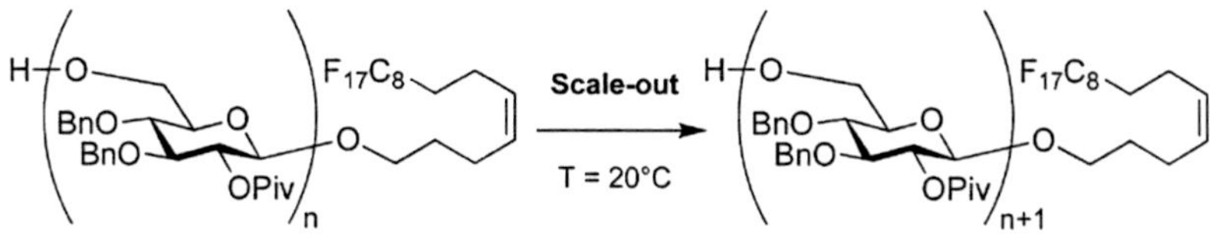

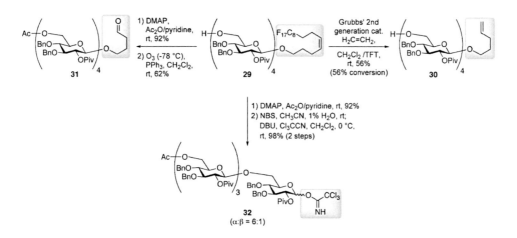

nucleophile	scale (mmol)	solvent	time [s]	product	yield
n = 0, **25**	0.214	TFT	30	**26**	99%
n = 1, **26**	0.176	CH₂Cl₂	20	**27**	97%
n = 2, **27**	0.102	CH₂Cl₂	60	**28**	90%
n = 3, **28**	0.073	CH₂Cl₂	60	**29**	95%

TFT: Trifluorotoluene

Scheme 5. Scale-out of the established coupling procedures [3e].

Scheme 6. Functionalization of the perfluorinated linker after successful assembly of the homotetramer **29** [3e].

β-Amino acid oligomers (β-peptides) present a unique class of peptides. In contrast to their natural α-amino acid derived counterparts, β-peptides show remarkable metabolic stability and are therefore of increasing interest for the pharmaceutical industry and the treatment of various diseases. Nevertheless, β-peptides require relatively few residues to form secondary structures like turns, helices or sheets [18], which usually leads to a lack of solubility in commonly employed solvents. Therefore, the synthesis of β-peptide structures is severely hampered [19]. Usually, high coupling temperatures are applied to circumvent precipitation of the peptides during the reaction, but the problem of low solubility remains unsolved for the required purification steps.

We investigated the application of a silicon-glass microfluidic reactor to the assembly of oligo-β-peptides (Scheme 7) [3c]. The microfluidic device thereby not only allowed for quick screening of reaction conditions and the controlled heating of the reaction mixture to unconventionally high temperatures, but also the procurement of synthetically useful amounts of peptides. The attachment of a perfluorinated linker again allowed for facile purification of the reaction mixtures by FSPE [17] to yield the desired oligopeptides **35**, **37** and **39** in excellent yields and short reaction times. Notably, the high reaction temperatures of up to 120 °C did not affect deprotection of the *tert*-butyloxycarbonyl (Boc) protected peptides, and highly reactive β^2-and β^3-homoaminoacid fluorides were employed for the β-peptide couplings. Furthermore, all possible β-peptide linkages were successfully installed, even the sterically most demanding β^3-β^2 linkage. Including global deprotection and cleavage from the perfluorinated linker the microreactor approach significantly increased the isolated overall yield of desired tetrapeptide **39** compared to solid phase or solution phase approaches (Scheme 7) [3c].

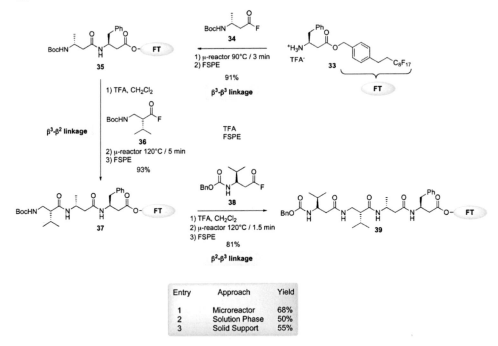

Entry	Approach	Yield
1	Microreactor	68%
2	Solution Phase	50%
3	Solid Support	55%

Scheme 7. Synthesis of a β-tetrapeptide in continuous-flow [3c].

CHEMICAL REACTIONS: EVALUATION OF MICROREACTORS FOR HAZARDOUS CHEMICAL TRANSFORMATIONS AND SCALE-UP

More recently, we explored the applicability of microreactor technology for hazardous synthetic transformations that are unattractive, difficult or even impossible to be performed using batch procedures, especially on large scale. With the aim to speed up the transfer from development stage to production scales in chemical companies for synthetically useful transformations, we investigated reactions such as $AlMe_3$ mediated amide bond formations [3f], deoxyfluorinations using diethylaminosulfur trifluoride (DAST) [3 g], radical defunctionalisations and hydrosilylations in commercially available glass and tube microreactors (Fig. 4) [3 h, 6].

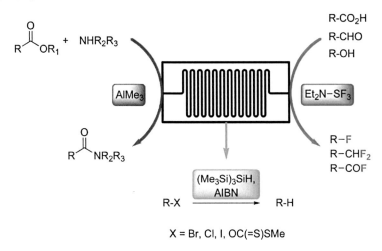

Figure 4. Microreactor based approaches toward hazardous reactions and scale-up chemistry [3f, g, h].

Amide bond formations are frequently used transformations in chemistry laboratories and generally require, starting from esters, a three step procedure including hydrolysis, activation and treatment with an amine to be synthesized (Scheme 8). Aluminium-mediated amide bond formations resembling direct Weinreb-amidation methods are, even though the reaction conditions render the transformation highly tolerant towards further functional groups in the molecule, less commonly applied due to the difficulties in safe handling of trimethylaluminium ($AlMe_3$). In addition, the aluminium-amide intermediates are unstable at elevated temperatures and are known to result in severe exotherms at room temperature. We investigated the applicability of microreactor technology for aluminium-mediated amide bond formation to develop a general protocol for the safe and rapid production of useful amounts of material ($1-2$ mol/day). A huge variety of primary and secondary amines as well as carboxylic acids derivatives were transformed into their corresponding amides in just two minutes reaction time (Scheme 9). Notably, installing a simple backpressure regulator allowed for superheating of the solvents while still allowing for the reaction to proceed chemoselectively (amide **60**) and to tolerate functional groups such as carbon-carbon double bonds (54, 56), hydroxyl groups (**55, 57**) or carbocycles (**58, 59**).

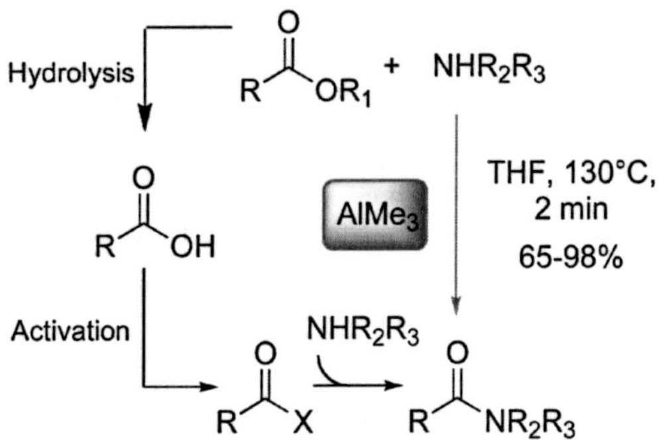

Scheme 8. General pathway for the formation of amides [3f].

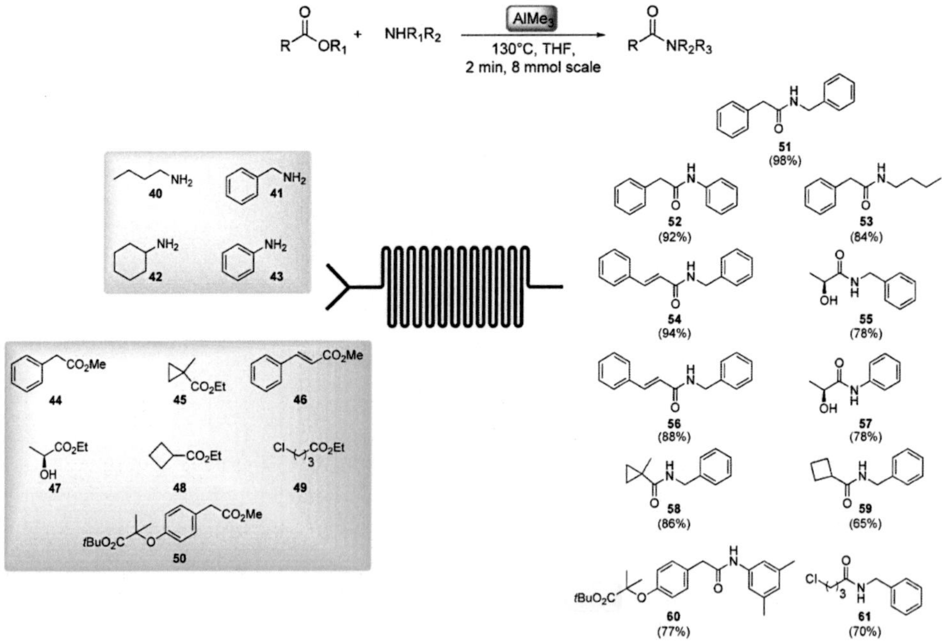

Scheme 9. Selected examples for AlMe$_3$ mediated direct amide bond formation [3f].

The developed general synthetic protocol was further applied to the continuous synthesis of Rimonabant® (Scheme 10) [3f]. Rimonabant® is an anti-obesity drug by Sanofi-Aventis® that acts as a central cannabinoid receptor antagonist and is approved in Europe [20]. The entire synthesis was performed in continuous-flow, starting from aromatic ketone **62** and diethyloxalate **63** to form diketone **64** in 70% over 10 min. Formation of pyrazole **66** was established in AcOH at 125 °C and a reaction time of 15 min. As the last synthetic transformation, the AlMe$_3$ mediated amide bond formation was performed to yield Rimonabant® **68** in an overall yield of 49% (Scheme 10) [3f].

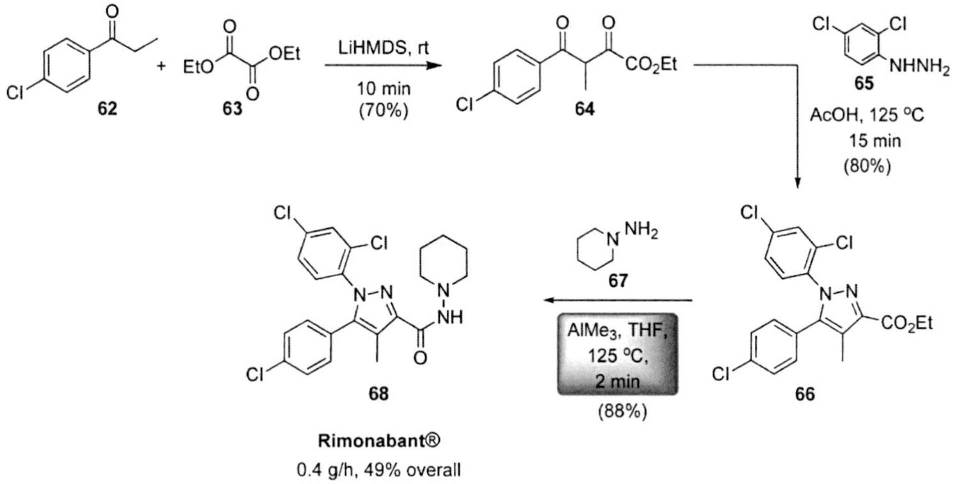

Scheme 10. Synthesis of Rimonabant® **68** in continuous-flow with AlMe$_3$ mediated amide bond formation as the key step [3f].

Fluorinated biologically active organic compounds are of great interest in pharmaceutical industry due to their unique biochemical and physical properties. Besides the generation of fluorinated molecules by nucleophilic substitutions starting from halides and HF or KF [21], or electrophilic fluorinations of β-diketones via an enole pathway [22], deoxyfluorinations using deoxofluor or DAST are convenient reaction pathways due to the ready availability of unprotected hydroxyl functionalities in organic molecules [3 g, 23]. Nevertheless, the synthesis of fluorinated drug candidates, their precursors or organic molecules in general via deoxyfluorinations is severely hampered due to safety concerns. The most commonly used reagent DAST detonates at 90 °C, forms HF after contact with moisture or water which makes the use of special equipment necessary and is ideally avoided on large scale. We were interested in investigating DAST mediated deoxyfluorinations of alcohols, aldehydes, lactols and carboxylic acids in a continuous-flow PTFE tube reactor. Installation of a backpressure regulator allowed for superheating of THF close to the detonation point of DAST, and an *in-situ* quench with aqueous NaHCO$_3$ immediately destroyed remaining DAST or formed HF (Fig. 5) [3 g].

R−CO2H
R−CHO
R−OH
16 minutes
R−F
R−CF2H
R−COF
Et2N−SF3
in situ quench

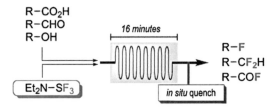

Figure 5. Overview of the DAST mediated deoxyfluorinations in continuous-flow [3 g].

Various primary and secondary alcohols underwent transformation into the corresponding fluorinated building block in high yields (Scheme 11). Consistent with the overall effort of our group to facilitate oligosaccharide assembly, electron rich and electron deficient glucosyl fluorides **74** and **75** were rapidly obtained in high yields from the corresponding lactols and did not require further purification before being used as glycosylating agents.

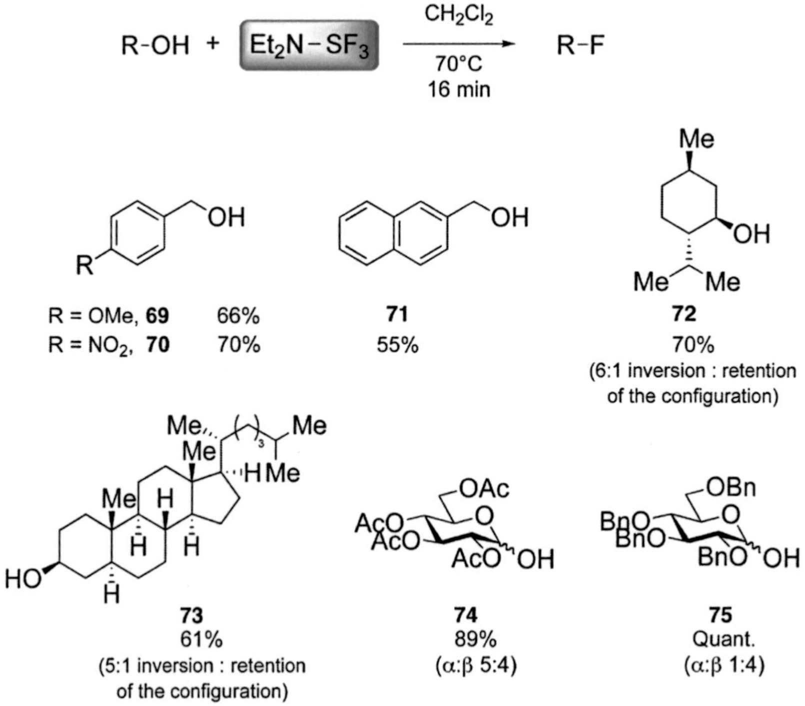

R-OH + Et2N-SF3
CH2Cl2
70°C
16 min
R-F
R = OMe, 69 66%
R = NO2, 70 70%
71
55%
72
70%
(6:1 inversion : retention of the configuration)
73
61%
(5:1 inversion : retention of the configuration)
74
89%
(α:β 5:4)
75
Quant.
(α:β 1:4)

Scheme 11. Deoxyfluorinations of various alcohols and lactols [3 g].

In a further attempt to rapidly generate reactive intermediates, aromatic and aliphatic carboxylic acids were transformed into their corresponding acid fluorides (Scheme 12). Particularly impressive is the selective formation of the acid fluorides in case enolizable α-
protons are present (**78, 79, 80**), no α-fluorination was observed. Furthermore, acid chloride
81 was selectively converted into the corresponding acid fluoride.

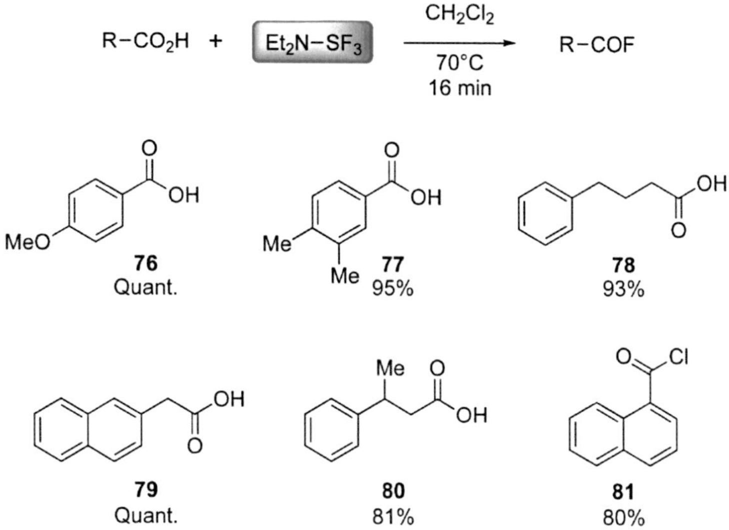

Scheme 12. Deoxyfluorinations of carboxylic acids and acid fluorides [3 g].

To generate difluoromethylene moieties, electron rich, electron deficient and sterically demanding aromatic aldehydes as well as aliphatic aldehydes underwent deoxyfluorination to
yield the analogous fluorinated derivative in high yields (Scheme 13). Remarkably, ketone
87 as a sterically and electronically more challenging substrate, was transformed into the
corresponding difluoromethylene derivative in 40% yield [3 g, 24].

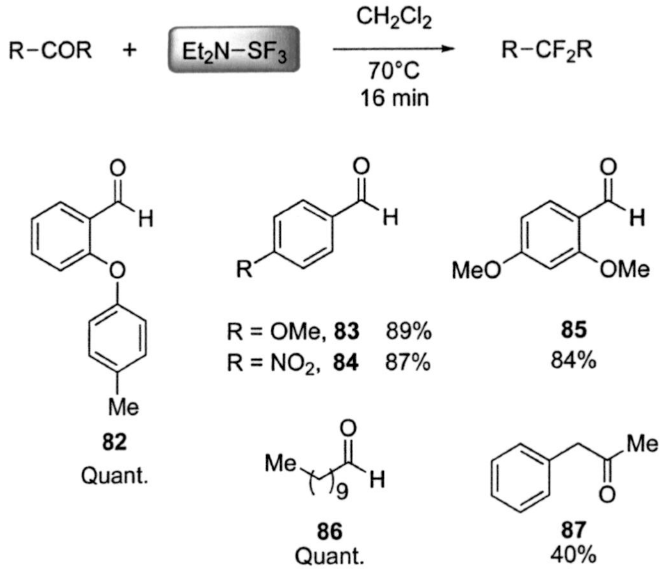

CH2Cl2
R−COR
Et2N−SF3
R−CF2R
70°C
16 min
R = OMe, 83 89%
R = NO2, 84 87%
85
84%
82
Quant.
86
Quant.
87
40%

Scheme 13. Synthesis of difluoromethylene moieties *via* DAST mediated deoxyfluorinations [3 g].

The latest investigations in our laboratories explored free radical mediated reactions such as dehalogenations, Barton-McCombie type deoxygenations and hydrosilylations in continuous-flow (Fig. 6) [3 h].

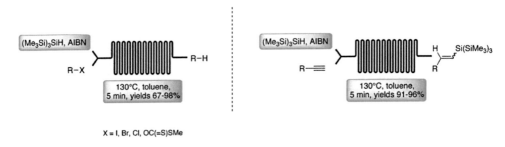

(Me3Si)3SiH, AIBN
R−X
R−H
130°C, toluene,
5 min, yields 67-98%
(Me3Si)3SiH, AIBN
Si(SiMe3)3
130°C, toluene,
5 min, yields 91-96%
X = I, Br, Cl, OC(=S)SMe

Figure 6. Free radical based transformations performed in glass microreactors [3 h].

Deoxygenations and dehalogenations are important synthetic transformations since they are highly versatile reactions and tolerate a huge variety of further functional groups in the molecules. Unfortunately, the concentration of free radicals in the reaction mixture often influences the selective outcome of the reaction strongly. Careful control of the reaction temperature is required to circumvent thermal runaways. Investigating reactions in a continuous-flow glass microreactor using tris(trimethylsilyl)silane (TTMSS) as a non-toxic variant of the generally applied tin reagent Bu_3SnH resulted in high yielding dehalogenations and deoxygenations at a reaction temperature of 130 °C (superheated toluene, established by a backpressure regulator) and a reaction time of five minutes (Scheme 14). Notably, taking advantage of the controlled reaction performance in the microreactor, the xanthate moiety of the difunctionalized dodecane **96** was removed chemoselectively [3 h].

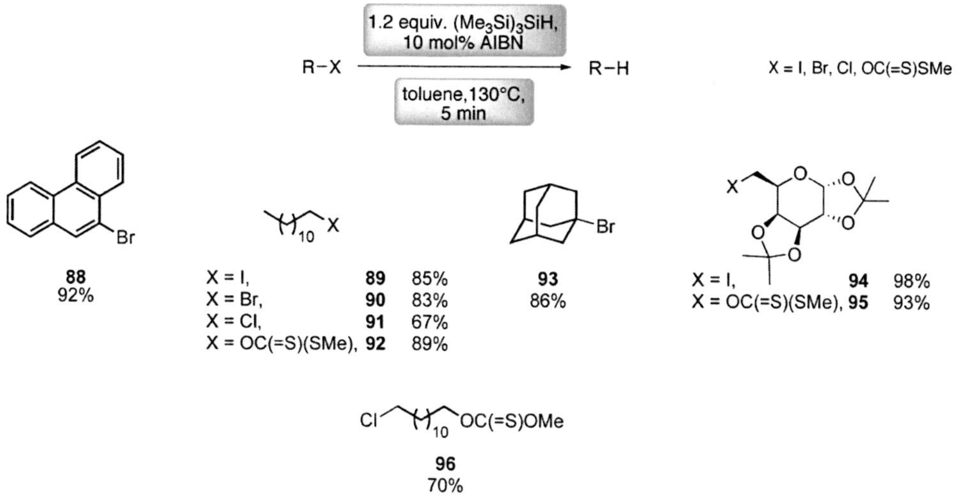

Scheme 14. Radical mediated dehalogenations and deoxygenations in continuous-flow [3 h].

Hydrosilylations, the addition of Si-H bonds across unsaturated carbon-carbon bonds, are the most important synthetic transformation for the generation of organosilicon compounds. Especially tris(trimethylsilyl)silane TTMSS adds highly regioselectively to various alkynes via a free-radical chain mechanism to form vinyl silanes [25]. We envisaged that the rapid heat transfer in microreactors would effect the E/Z selectivity of the resulting silanes while significantly reducing the reaction time (Scheme 15).

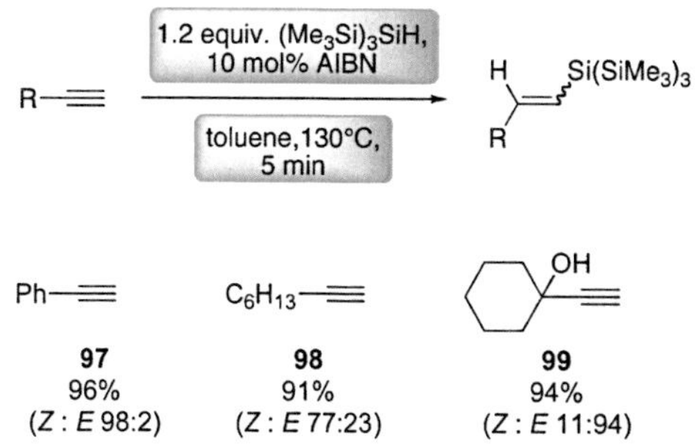

Scheme 15. Hydrosilylations of alkynes performed in a glass microreactor [3 h].

Again applying reaction conditions of 130 °C in toluene and a reaction time of five minutes yielded the desired anti-Markovnikov products in excellent yields and good stereo-selectivities. The microreactor approach further furnished the hydrosilylation product of phenylacetylene **97** in greater yield (96% *vs.* 88%) and superior *Z/E* ratio (98:2 *vs.* 84:16) compared to traditional batch strategies [3 h, 25].

Conclusion

Many chemical transformations can be performed in a faster, safer and cleaner manner when performed in microfluidic devices. The down-scaling of reaction volumes in microreactors offers the more precise control of the reaction conditions (temperature, time, mixing) and the use of minimal amounts of precious compounds to rapidly screen a variety of conditions, generating a wealth of information on reaction kinetics and pathways. Additionally, microreactors present the opportunity to apply reaction conditions that are inaccessible using conventional laboratory equipment, such as super heated solvents, and safely performing reactions in explosive regimes. Even though numerous impressive examples on the application of microreactor technology in synthetic chemistry have been reported, some major drawbacks associated with this technique remain: the incompatibility of reactors with solid reagents that cannot be used for wall-coatings or be supported on linkers, the sensitivity to precipitations and the useful applicability mainly to fast reactions. Additionally, the efficient *on-line* analysis of reaction mixtures in a high throughput format represents an outstanding issue. The overall concept of highly efficient continuous-flow microreactor techniques requires further improvements before it will be applied as a standard tool in chemical laboratories.

References

[1]   For recent reviews and books on microchemical technology, see: (**a**) Ehrfeld, W., Hessel, V., Löwe, H. (2000) *Microreactors: New Technology for Modern Chemistry*, Wiley-VCH, Weinheim. (**b**) Fletcher, P.D.I., Haswell, S.J., Pombo-Villar, E., Warrington, B.H., Watts, P., Wong, S.Y.F., Zhang, X.L. (2002) *Tetrahedron*, **58**:4735. (**c**) Hessel, V., Hardt, S., Löwe, H. (2004) *Chemical Micro Process Engineering*, Wiley-VCH, Weinheim. (**d**) Jähnisch, K., Hessel, V., Löwe, H., Baerns, M. (2004) *Angew. Chem. Int. Ed.* **43**:406. (**e**) Thayer, A.M. (2005) *Chem. Eng. News*, **83**:43. (**f**) Watts, P., Haswell, S.J. (2005) *Chem. Soc. Rev.* **34**:235. (**g**) Geyer, K., Codée, J.D.C. Seeberger, P.H. (2006) *Chem. Eur. J.* **12**:8434. (**h**) Brivio, M., Verboom, W., Reinhoudt, D.N. (2006) *Lab Chip*, **6**:329. (**i**) Mason, B.P., Price, K.E., Steinbacher, J.L., Bogdan, A.R., McQuade, D.T. (2007) *Chem. Rev.*, **107**:2300. (**j**) Ahmed-Omer, B., Brandt, J.C., Wirth, T. (2007) *Org. Biomol. Chem.* **5**:733. (**k**) Fukuyama, T., Rahman, T., Sato, M., Ryu, I. (2008) *Synlett.* **2**:151. (**l**) Wiles, C., Watts, P. (2008). *Eur. J. Org. Chem.* **10**:1655.

[2]   For recent publications of microreactors in organic synthesis see: (**a**) Usutani, H., Tomida, Y., Nagaki, A., Okamoto, H., Nokami, Y., Yoshida, J. (2007) *J. Am. Chem. Soc.* **129**:3046. (**b**) Hartung, A., Keane, M.A., Kraft, A. (2007) *J. Org. Chem.* **72**:10235. (**c**) Bula, W.P., Verboom, W., Reinhoudt, D.N., Gardeniers, H.J. (2007) *Lab Chip* **7**:1717. (**d**) Sahoo, H.R., Kralj, J.G., Jensen, K.F. (2007) *Angew. Chem. Int. Ed.* **46**:5704. (**e**) Tanaka, K., Motomatsu, S., Koyama, K., Tanaka, S.I., Fukase, K. (2007) *Org. Lett.* **9**:299. (**f**) Trapp, O., Weber, S.K., Bauch, S., Hofstadt, W. (2007) *Angew. Chem. Int. Ed.* **46**:7307. (**g**) Fukuyama, T., Kobayashi, M., Rahman, T., Kamata, N., Ryu, I. (2008) *Org. Lett.* 10:433. (**h**) Baumna, M., Baxendale, I.R., Ley, S.V., Nikbin, N., Smith, C.D., Tierney, J.P. (2008) *Org. Biomol. Chem.* **6**:1577. (**i**) Nagaki, A., Takabayashi, N., Tomida, Y., Yoshida, J. (2008) *Org. Lett.* DOI:10.1021/ol8015572.

[3]   For recent examples of microreactor-based transformations from this laboratory see (**a**) Snyder, D.A., Noti, C., Seeberger, P.H., Schael, F., Bieber, T., Rimmel, G., Ehrfeld, W. (2005) *Helv. Chim. Acta* **88**:1. (**b**) Ratner, D.M., Murphy, E.R., Jhunj-hunwala, M., Snyder, D.A., Jensen, K.F., Seeberger, P.H. (2005) *Chem. Comm.* **5**:578. (**c**) Flögel, O., Codée, J.D.C., Seebach, D., Seeberger, P.H. (2006) *Angew. Chem. Int. Ed.* **45**:7000. (**d**) Geyer, K., Seeberger, P.H. (2007) *Helv. Chim. Acta* **90**:395. (**e**) Carrel, F.R., Geyer, K., Codée, J.D.C., Seeberger, P.H. (2007) *Org. Lett.* **9**:2285. (**f**) Gustafsson, T., Pontén, F., Seeberger, P.H. (2008) *Chem.Commun.* **10**:1100. (**g**) Gustafsson, T., Gilmour, R., Seeberger, P.H. (2008) *Chem. Commun.* 3022. (**h**) Odedram, A., Geyer, K., Gustafsson, T., Gilmour, R., Seeberger, P.H. (2008) *Chem. Commun.* 3025.

[4]   Reagent introduction and feed can occur via electroosmotic flow, hydrodynamic pumping (HPLC pumps, syringe pumps) or capillary flow.

[5]   A wide variety of other microreactors is available from other manufacturers and suppliers, *e.g.*: Institute for Microtechnology Mainz (IMM), Fraunhofer Alliance for Modular Microreaction Systems (FAMOS), the Little Things Factory (LTF) in Ilmenau, the New Jersey Centre for MicroChemical Systems (NJCMCS), the Micro-Chemical Process Technology Research Association (MCPT), or Sigma-Aldrich. The selection presented here is by no means exhaustive and only serves to indicate the diversity in systems developed to date.

[6]   For further information visit the webpage: http://www.syrris.com/.

[7]   For further information visit the webpage: http://www.ehrfeld.com.

[8]   For further information visit the webpage: http://www.cpc-net.com/cytosls.shtml.

[9]   For further information visit the webpage: http://www.micronit.com/

[10]   For further information visit the webpage: http://www.vapourtec.co.uk/

[11]   For recent reports on biological and biochemical applications see: (a) Hansen, C.L., Classen, S., Berger, J.M., Quake, S.R. (2006) *J. Am. Chem. Soc.* 3142. (b) Wang, B., Zhao, Q., Wang, F., Gao, C. (2006) *Angew. Chem. Int. Ed.* **45**:1560. (c) Duan, J., Sun, L., Liang, Z., Zhang, J., Wang, H., Zhang, L., Zhang, W., Zhang, Y. (2006) *J. Chromatogr. A* **1–2**:165. (d) Utsumi, Y, Hitaka, Y., Matsui, K., Takeo, M., Negoro, S., Ukita, S. (2007) *Microsystem Technologies* **13**:425. (e) Chen, L., West, J., Auroux, P.A., Manz, A., Day, P.J. (2007) *Anal. Chem.* **79**:9185. (f) Liu, Y., Qu, H., Xue, Y., Wu, Z., Yang, P., Liu, B. (2007) *Proteomics*, **7**:1373. (g) Le Nel, A., Minc, N., Smadja, C., Slovakova, M., Bilkovam, Z., Peyrin, J.M., Viovy, J.L., Taverna, M (2008) *Lab Chip* **8**:294. (h) Yamamoto, T., Hino, M., Kakuhata, R., Nojima, T., Shinohara, Y., Baba, Y., Fujii, T. (2008) *Anal. Sci.* **24**:234.

[12]   For reports on larger scale productions using stainless steel microreactors see: (a) Taghavi-Moghadam, S., Kleemann, A., Golbig, K.G. (2001) *Org. Process Res. Dev.*, **5**:652. (b) Panke, G., Schwalbe, T., Stirner, W., Taghavi-Moghadam, S., Wille, G. (2003) *Synthesis* **18**:2827. (c) Zhang, X.N., Stefanick, S., Villani, F.J. (2004) *Org. Process Res. Dev.* **8**:455. (d) Liu, S.F., Fukuyama, T., Sato, M., Ryu, I. (2004) *Org. Process Res. Dev.* **8**:477. (e) Acke, D.R.J, Stevens, C.V. (2006) *Org. Process Res. Dev.* **10**:417. (f) Wheeler, R.C., Benali, O., Deal, M., Farrant, E., MacDonald, S.J.F., Warrington, B.H. (2007) *Org. Process Res. Dev.* **11**:704.

[13]   For reports on the application of glass microreactors see: (a) Nikbin, N., Watts, P. (2004) *Org. Process Res. Dev* **8**:942. (b) Wiles, C., Watts, P., Haswell, S.J. (2007) *Chem. Commun.* **9**:966. (c) Wiles, C., Watts, P., Haswell, S.J. (2007) *Lab Chip* **7**:322. (d) Hornung C.H., Mackley, M.R., Baxendale, I.R., Ley, S.V. (2007) *Org. Process Res. Dev.* **11**:399. (e) Smith, C.D., Baxendale, I.R., Tranmer, G.K., Baumann, M., Smith, S.C., Lewthwaite, R.A., Ley, S.V. (2007) *Org. Biomol. Chem.* **10**:1562. (f) Smith, C.D., Baxendale, I.R., Lanners, S., Hayward, J.J., Smith, S.C., Ley, S.V. (2007) *Org. Biomol. Chem.* **10**:1559.

[14]   For recent reviews and reports on photochemical reactions using microflow see (a) Fukuyama, T., Hino, Y., Kamata, N., Ryu, I. (2004) *Chem. Lett.* **33**:1430. (b) Maeda, H., Mukae, H., Mizuno, K. (2005) *Chem. Lett.* **34**:66. (c) Maeda, H., Matsukawa, N., Shirai, K., Mizuno, K. (2005) *Tetrahedron Lett.* **46**:3057. (d) Hook, B.D.A., Dohle, W., Hirst, P.R., Pickworth, M., Berry, M.B., Booker-Milburn, L.I. (2005) *J. Org. Chem.* **70**:7558. (e) Matsushita, Y., Kumada, S., Wakabayashi, K., Sakeda, K., Ichimura, T. (2006) *Chem. Lett.* **35**:410. (f) Lainchbury, M.D., Medley, M.I., Taylor, P.M., Hirst, P., Dohle, W., Booker-Milburn, L.I. (2008) *J. Org. Chem.* DOI: 10.1021/jo801108 h.

[15]   For selected reviews and reports on silicon based microreactors see (a) Jensen, K.F. (2001) *Chem. Eng. Sci.* **56**:293. (b) Yen, B.K.H., Gunther, A., Schmidt, M.A., Jensen, K.F., Bawendi, M.G. (2005) *Angew. Chem. Int. Ed.* **44**:5447. (c) Murphy, E.R.,

Martinelli, J.R., Zaborenko, N., Buchwald, S.L., Jensen, K.F. (2007) *Angew. Chem. Int. Ed.* **46**:1734. (**d**) Kralj, J.G., Sahoo, H.R., Jensen, K.F. (2007) *Lab Chip* **7**:256. (**e**) Inoue, T., Schmidt, M.A., Jensen, K.F. (2007) *Ind. Eng. Chem. Res.* **46**:1153.

[16] (**a**) Ravida, A., Liu, X.Y., Kovacs, L., Seeberger, P.H. (2006) *Org. Lett.* **8**:1815. (**b**) Liu, X.Y., Stocker, B.L., Seeberger, P.H. (2006) *J. Am. Chem. Soc.* **128**:3638.

[17] (**a**) Curran, D.P. (2001) *Synlett.* **9**:1488. (**b**) Curran, D.P., Luo, Z.Y. (1999) *J. Am. Chem. Soc.* **121**:9069. (**c**) Zhang, W. (2004) *Chem. Rev.* **104**:2531.

[18] (**a**) Murray, J.K., Gellman, S.H. (2005) *Org. Lett.* **7**:1517. (**b**) Lelais, G., Seebach, D., Jaun, B., Mathad, R.I., Flögel, O., Rossi, F., Campo, M., Wortmann, A. (2006) *Helv. Chim. Acta* **89**:361.

[19] Erdélyi, M., Gogoll, A. (2002) *Synthesis* 1592–1596.

[20] Monnier, O., Coquerel, G., Fours, B., Duplaa, H., Ochsenbein, P., PTC Patent, PCT/ FR2007/000201, 05.02.2007.

[21] For selected examples on nucleophilic substitutions using HF or fluorine salts, see (**a**) Kim, D.W., Song, C.E., Chi, D.Y. (2003) *J. Org. Chem.* **68**:4281. (**b**) Inagaki, T., Fukuhara, T., Hara, S. (2003) *Synthesis* 1157.

[22] For electrophilic fluorinations see for example Xiao, J., Shreeve, J.M. (2005) *J. Fluorine Chem.* **126**:473.

[23] For selected examples on fluorinations using DAST or deoxofluor see (**a**) Lal, G.S., Pez, G.P., Pesaresi, R.J., Prozonic, F.M., Cheng, H. (1999) *J. Org. Chem.* **64**:7048. (**b**) White, J.M., Tunoori, A.R., Turunen, B.J., Georg, G.I. (2004) *J. Org. Chem.* **69**:2573.

[24] These finding are consistent with the literature, see also Negi, D.S., Köppling, L., Lovis, K., Abdallah, R., Geisler, J., Budde, U. (2008) *Org. Process Res. Dev.* **12**:345.

[25] Kopping, B., Chatgilialoglu, C., Zehnder, M., Giese, B. (1992) *J. Org. Chem.* **57**:3994.

CHEMICAL BIOLOGY WITH ORGANOMETALLICS

ERIC MEGGERS

Fachbereich Chemie, Philipps-Universität Marburg,
Hans-Meerwein-Straße, 35043 Marburg, Germany

E-Mail: meggers@chemie.uni-marburg.de

Received: 7th August 2008 / Published: 16th March 2009

ABSTRACT

Inorganic and organometallic moieties are explored as structural scaffolds for the design of biologically active compounds. In this strategy, a metal center plays a structural role by organizing the organic ligands in the three-dimensional receptor space. It is hypothesized that this approach allows to access unexplored chemical space, thus giving new opportunities for the design of small molecules with unprecedented biological properties. Along these lines, our group pioneered the design of organometallic scaffolds for the inhibition of protein kinases. These compounds are formally derived from the class of ATP-competitive indolocarbazole alkaloids and allow to access novel structures with very defined and rigid shapes in an economical fashion, having resulted in the discovery of picomolar inhibitors for protein kinases with impressive selectivity profiles. This article summarizes the most important results over the last few years.

INTRODUCTION

The identification of compounds with novel and defined biological functions is of high importance for the development of small-molecule pharmaceuticals and as tools ("molecular probes") for the investigation of biological processes. Nowadays, established technologies such as combinatorial chemistry and high-throughput screening allow to synthesize and identify initial lead structures in a rapid fashion and, in this respect, the chemical diversity of scaffolds is a key factor for success. The stark majority of synthetic efforts is focused on purely organic compounds. Surprisingly, a recent publication revealed that the diversity of organic molecules is however quite limited: half of the 24 million organic compounds registered in the Chemical Abstract Service (CAS) Registry can be classified with only 143 scaffolds [1]. Similarly, the diversity of topological shapes of known organic drugs is also extremely low: a report on the analysis of the Comprehensive Medicinal Chemistry (CMC) database demonstrated that out of more than 5000 compounds analyzed, half of the known drugs fall into only 32 shape categories [2]. Most of the 32 frameworks contain at least two six-membered rings linked or fused together. Thus, it can be concluded that current efforts to obtain biologically active compounds draw from a limited set of structural scaffolds. Strategies to increase the molecular structural diversity are therefore demanded and promise to provide opportunities for the discovery of compounds with novel and unprecedented properties.

METAL COMPLEXES AS STRUCTURAL SCAFFOLDS

A few years ago, our group hypothesized that structural diversity might be increased by complementing organic elements with single metal atoms and to explore the opportunities of metals to help building small compounds with defined three dimensional structures [3]. Transition metals appear especially appealing for this purpose since they can support a multitude of coordination numbers and geometries which reach beyond the linear (sp-hybridization), trigonal planar (sp^2-hybridization), and tetrahedral (sp^3-hybridization) binding geometries of carbon (Fig. 1). For example, it is intriguing that an octahedral center with six different substituents is capable of forming 30 stereoisomers compared to just two for an asymmetric tetrahedral carbon. Thus, by increasing the number of substituents from four (tetrahedral center) to six (octahedral center), the ability of the center to organize substituents in the three dimensional space increases substantially. In addition, using a hexavalent center may provide new synthetic opportunities for accessing globular shapes by building structures from a single center in six different directions.

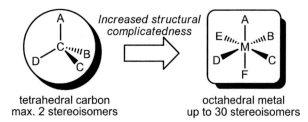

tetrahedral carbon
max. 2 stereoisomers

octahedral metal
up to 30 stereoisomers

Figure 1. Structural opportunities with an octahedral metal center.

RUTHENIUM AS A PRIVILEDGED METAL

Current efforts in our laboratory are concentrating predominately on ruthenium. In our opinion, ruthenium possesses a combination of properties that render it ideal for the design of chemically inert bioactive organometallic species. Main reasons are a high kinetical inertness of coordinative bonds to ruthenium if certain basic design principles are considered (all here discussed compounds are air-stable, stable in water, and can withstand millimolar concentrations of thiols and thus should stay intact within the biological environment), the availability of inexpensive metal precursors (currently US$ 20/g for RuCl₃), and a predictable and established synthetic chemistry. A significant concern in using metal-containing compounds is the potential toxicity of the metal. In this respect, it is our opinion that a metal complex that has a substitutionally inert coordination sphere should not display any "metal-specific" toxicity itself and the biological properties of the metal complex should be instead determined by the entire entity. However, toxic effects due to metabolic conversion of coordinated ligands followed by the exposure of the metal to biological ligands cannot be excluded in individual cases. It is therefore desirable that the metal itself or derived metal salts have a low toxicity. In this respect, data from phase I clinical trials of two (substitutionally labile) ruthenium anticancer drugs, KP1019 and NAMI-A, are encouraging since these compounds have proven to be much less toxic compared to the anticancer drug cisplatin [4, 5].

STAUROSPORINE AS A LEAD STRUCTURE

We selected the natural product staurosporine as a structural inspiration (lead structure) for the design of ATP-competitive organometallic kinase inhibitors **1** (Figs. 2 and 3) [6]. Staurosporine is a member of the class of indolocarbazole alkaloids, many of which are potent ATP-competitive protein kinase inhibitors. This family of inhibitors shares the indolo[2,3-a]carbazole moieties **2** (lactam form) or **3** (imide form, arcyriaflavin A) which bind to the adenine binding site by establishing two hydrogen bonds to the backbone of the hinge between the N-terminal and C-terminal kinase domain (Figs. 2b and 3). Staurosporine adopts a defined globular structure with the carbohydrate moiety being oriented perpendicular to the plane of the indolocarbazole heterocycle (Fig. 3). The indolocarbazole moiety occupies the hydrophobic adenine binding cleft with the lactam group mimicking the hydrogen bonding pattern of the adenine base, and the carbohydrate moiety forming hydrophobic contacts and hydrogen bonds within the globular ribose binding site (Fig. 2). Thus, staurosporine closely matches the shape of the ATP-binding site which makes it a highly potent albeit unselective inhibitor for many protein kinases.

In order to match the overall shape of the ATP-binding site of protein kinases in a fashion similar to staurosporine, but with less synthetic effort and more extended structural options, we replaced the indolocarbazole alkaloid structure with simple metal complexes in which the main features of the indolocarbazole heterocycle **3** are retained in a metal-chelating pyrido-carbazole **4** (Fig. 3). This ligand **4** can serve as a bidentate ligand for ruthenium complexes of type **1**. Additional ligands in the coordination sphere of the metal can now substitute for the carbohydrate moiety of staurosporine, with the metal center serving as a "glue" to unite all of the parts. This approach has resulted in the successful design of nanomolar and even picomolar protein kinase inhibitors.

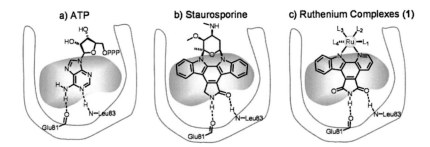

Figure 2. Schematic representation of the ATP-binding pocket of cyclin dependent protein kinase 2 (CDK2) in complex with a) ATP, b) staurosporine, and c) indolocarbazole-derived ruthenium complexes. Both ATP and staurosporine form hydrogen bonds with the backbone amide groups of glutamate 81 and leucine 83. The green area indicates a patch of high hydrophobicity. Adapted from ref. [7].

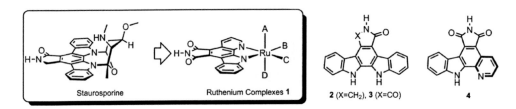

Figure 3. Designing ruthenium complexes of scaffold **1** as protein kinase inhibitors by using the ATP-competitive natural product staurosporine as a lead for the overall three-dimensional shape. The chelating ligand **4** contains the main pharmacophore of the indolocarbazole heterocycles **2** and **3**.

PROOF-OF-PRINCIPLE: ORGANORUTHENIUM INHIBITORS FOR THE PROTEIN KINASES GSK-3 AND PIM-1

We recently identified the cyclopentadienyl-CO sandwich scaffold **5** to be a nanomolar inhibitor for the protein kinases GSK-3 and Pim-1 (Fig. 4). IC_{50} curves of racemic **5** and the corresponding pyridocarbazole ligand **4** against the α-isoform of GSK-3 (GSK-3α) are shown in Figure 4c. Accordingly, the pyridocarbazole ligand **4** itself is 15000-times less potent against GSK-3α compared to the entire complex **5**. Consequently, the activity of **5** requires the complete ruthenium ligand sphere. Methylation of **5** at the imide nitrogen (**Me5**) abolishes this activity completely. Figure 4c furthermore demonstrates that staurosporine is by an order of magnitude less potent for this kinase than **5** [8].

Complex **5** displays metal-centered chirality and potency as well as selectivity for GSK-3 or Pim-1 depend on the absolute configuration and additional substitutions at the periphery of this half-sandwich scaffold. For example, whereas the racemic mixtures of **5** and **6** are dual inhibitors for GSK-3 and Pim-1, (*S*)-**6**, having a additional OH-group at the indole, is very selective for Pim-1 with an IC_{50} of 0.22 nM at 100 μM ATP compared to 9 nM for GSK-3α (α-isoform) [9]. On the other hand, (R)-7, having an OH and Br-group at the indole and a methylester at the cyclopentadienyl moiety, displays an IC_{50} of 0.35 nM for GSK-3α compared to 35 nM for Pim-1. In fact, in a panel of 57 kinases, (R)-7 is highly selective for just GSK-3 [10].

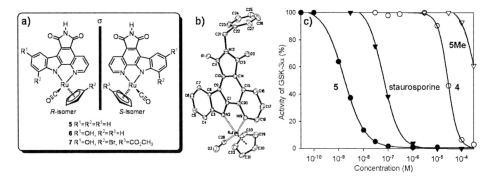

Figure 4. (a) The half-sandwich complexes **5-7** are potent inhibitors for the protein kinases GSK-3 and/or Pim-1. The half-sandwich compounds **5-7** are pseudo-tetrahedral and exist in two mirror-imaged configurations. The absolute configuration at the ruthenium center has been assigned according to the priority order of the ligands being η^5-C_5H_5 > pyridine [N(C, C, C)] > indole [N(C, C, lone pair)] > CO. (b) Crystal structure of N-benzylated **5** (ORTEP drawing with 35% probability thermal elipsoids). (c) IC_{50} curves with GSK-3α measured with 10 μM ATP. **5Me** is the N-methylated derivative of **5**.

COCRYSTAL STRUCTURES OF PROTEIN KINASES WITH ORGANOMETALLICS

In collaboration with Stefan Knapp (Oxford University) and later independently in our own laboratory, we obtained cocrystal structures of ruthenium complexes bound to Pim-1 [9]. For example, Figure 5 shows a cocrystal structure of Pim-1 with (S)-6. The structure demonstrates that the ruthenium complex (S)-6 binds to the ATP binding site and that the ruthenium centre does not undergo any direct interactions with the active site but has a purely structural role. Furthermore, superimposing cocrystallized (S)-6 and Pim-1 (PDB code 2BZI) with cocrystallized staurosporine and Pim-1 (PDB code 1YHS) reveals how closely the half-sandwich complex mimics the binding mode of staurosporine (Figs. 5b and 5c). The pyridocarbazole moiety of (S)-6 is nicely placed inside of hydrophobic pocket formed between residues from the N-terminal and C-terminal domain, mimicking the binding position of the indolocarbazole moiety of staurosporine, while the rest of the ruthenium complex occupies the binding site of the carbohydrate moiety of staurosporine. Curiously, the oxygen of the Ru-CO ligand is at a very close distance to Gly45 in the glycine-rich loop.

Currently, five structures of ruthenium half-sandwich complexes with protein kinases are available in the protein databank, four with Pim-1 (2BZH, 2BZI, 2OI4, 2BZJ), and one with Pim-2 (2IWI). In addition, a cocrystal structure of an osmium complex with Pim-1 was published recently by us (3BWF) [11]. A structure of (R)-**6** with GSK-3β has been solved in our laboratory at 2.4 Å and will be published soon (G. E. Atilla-Gockumen, E. Meggers, unpublished results). The Marmorstein group (The Wistar Institute, Philadelphia, USA) obtained several structures by soaking ruthenium complexes into crystal structures of the lipid kinase PI3Kγ [12].

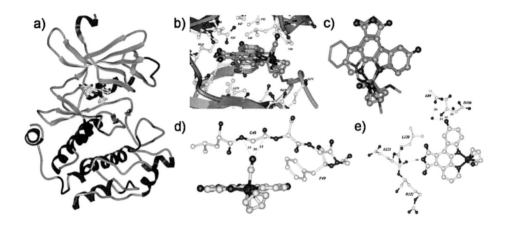

Figure 5. Cocrystal structure of (*S*)-**6** with Pim-1 at 1.9 Å (PDB code 2BZI). (**a**) The ruthenium complex occupies the ATP binding site of Pim-1. (**b**) and (**c**) Superimposed relative binding positions of (*S*)-**6** and staurosporine to Pim-1 (PDB code 1YHS). (**d**) and (**e**) Some important interactions.

CELLULAR DATA: INHIBITION OF GSK-3β BY A RUTHENIUM COMPLEX WITHIN MAMMALIAN CELLS

Over the last two year we have repeatedly demonstrated that organometallic GSK-3 inhibitors function inside of living mammalian cells [10, 13, 14]. GSK-3β is a negative regulator of the wnt-signaling pathway by phosphorylating β-catenin. Phosphorylated β-catenin itself is unstable due to rapid degradation by the proteasome. In the presence of a wnt signal, GSK-3β is inactivated, resulting in an accumulation of β-catenin in the cytoplasm, followed by a translocation into the nucleus where β-catenin serves as a transcriptional cofactor. Thus, inhibition of GSK-3β by pharmacological inhibitors or by wnt signaling leads to increased β-catenin levels and activation of wnt dependent transcription.

In order to probe the inhibition of GSK-3β within mammalian cells we routinely use Western blot analysis (qualitative detection of β-catenin upregulation), cellular β-catenin staining experiments (confirmation of β-catenin translocation into the nucleus), and a β-catenin reporter system for a quantitative analysis. For the latter we use human embryonic kidney cells (HEK-293OT) that have stably incorporated a Tcf-luciferase transcription reporter (OT-Luc cells). This transcription reporter generates luciferase in response to increased concentrations of β-catenin. An inhibition of GSK-3β can thus be determined by the luminescence signal upon addition of luciferin to the cell lysate. For example, exposure of OT-Luc cells to varying concentrations of the selective ruthenium GSK-3 inhibitor (R)-7 over a period of 24 hours yielded a strong upregulation of luciferase in the concentration window of 3 μM down to 10 nM (Figure 6a, compare also with (R,S)-5 in Figure 6b). Intriguingly, in contrast, the popular organic GSK-3 inhibitors 6-bromo-indirubin-3'-oxime (BIO) (8) and kenpaullone (9) (Figs. 6c, d) require micromolar concentrations for significant activities.

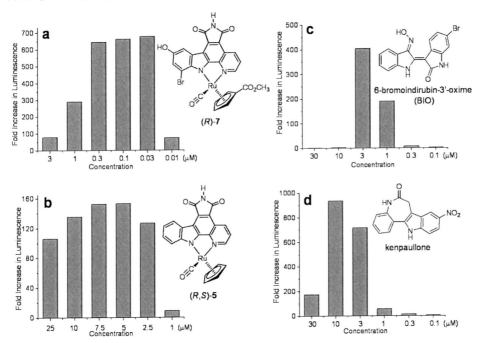

Figure 6. Cellular GSK-3β inhibition of (R)-7 and comparison with organic GSK-3 inhibitors. HEK-293OT cells transfected with a β-catenin-responsive luciferase reporter were treated with different concentrations of inhibitors for 24 hours. After cell lysis, the luminescence signal was measured after addition of luciferin.

It is noteworthy that organometallic compounds of this type can inhibit GSK-3β in frog embryos (demonstrated for (R,S)-6) and zebrafish embryos (demonstrated for (R)-7) and result in phenotypes that are related to the activation of the wnt pathway. For example, the exposure of zebrafish embryos to LiCl, a known GSK-3 inhibitor, causes a perturbed development of the head structure with a no-eye-phenotype, among others. Treatment of zebrafish embryos with (*R*)-**7** under analogous conditions, results in a similar phenotyp with a decrease of the head structure without eyes and a stunted and crooked tail. In addition, the yolk is enlarged and misshaped (Fig. 7c). However, zebrafish embryos which were instead treated under identical conditions with the mirror-imaged compound (*S*)-**7**, develop normal (Figs. 7a and 7b). These observations are consistent with an inhibition of GSK-3β by (*R*)-**7**, but not (*S*)-**7** and further demonstrate that the biological activity of (*R*)-**7** is determined by the overall three-dimensional structure of the ruthenium complex and not the ruthenium itself.

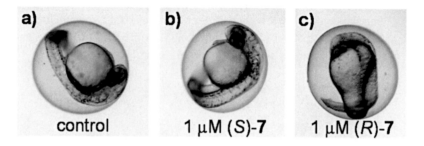

Figure 7. Exposure of zebrafish embryos to (**a**) DMSO (2%), (**b**) 1 μM (*S*)-**7**, and (**c**) 1 μM (*R*)-**7** (see Figure 4 for structures). The embryos were collected and maintained in E3 media at 28.5 °C, compounds added at 4 hours post fertilization (hpf), and the phenotypes were compared at 25 hpf.

In conclusion, organometallic (*R*)-**7** is superior as a molecular probe for the function of GSK-3β compared to established organic GSK-3β inhibitors with respect to binding affinity, cellular potency, and kinase selectivity.

COMBINATORIAL CHEMISTRY WITH RUTHENIUM COMPLEXES

We were interested in developing a method to rapidly assemble coordination spheres around the ruthenium center in a combinatorial fashion and for this we deviced a synthesis of complex **8** [15]. Although this compound bears four leaving groups, three acetonitriles and one chloride, it is remarkably stable at room temperature. For example, **18** can be purified on a regular silica gel column (CH₂Cl₂/MeOH) and is stable as a solid on the bench for a few days without decomposition. This behaviour is a demonstration of the kinetic inertness of the ruthenium pyridocarbazole scaffold. However, the acetonitrile ligands and the chloride can be replaced upon heating of **8** in the presence of other ligands at elevated temperatures of 70 – 110 °C (**8**→**9**) (Fig. 8a). Thus, **8** allows a rapid access to a diversity of novel structures **9** by ligand replacement chemistry in the final synthetic step.

Figure 8b shows a selection of octahedral ruthenium complexes **10-13** that we synthesized through this route. Compounds **10**, **11**, and **12** are interesting protein kinase inhibitors. For example, the racemic compound **11**, bearing a CO and 1,4,7-trithiacyclononane ligand, is a subnanomolar inhibitor for the protein kinase Pim-1 with an IC_{50} of 0.45 nM at 100 μM ATP. Intriguingly, the nature of this monodentate ligand has a dramatic effect on kinase inhibition. For example, substituting the CO for $P(OMe)_3$ (**13**) renders the compound completely inactive (Fig. 8b,c). Other ligands such as DMSO, cyanide, and ammonia also result in a significant decrease in potency for Pim-1 (data not shown). Interestingly, substituting the CO for an azide (**12**) yields a fairly potent MSK-1 inhibitor ($IC_{50} = 70$ nM).

The bar diagram in Figure 8c further demonstrates that the CO complex **11** is most potent against Pim-1 but also inhibits GSK-3α to a significant extent at 100 nM. Apparently, the CO ligand, which is in the plane perpendicular to the pyridocarbazole chelate, is an important pharmacophore both for Pim-1 and GSK-3, but not for most other kinases. The rest of the ligand sphere can then be used to render the complex either selective for Pim-1 or GSK-3. This is well illustrated by the compounds **10** and **11**. Whereas complex **11** significantly prefers Pim-1, the complex **10** is instead selective for GSK-3α with an IC_{50} of 8 nM at 100 μM ATP.

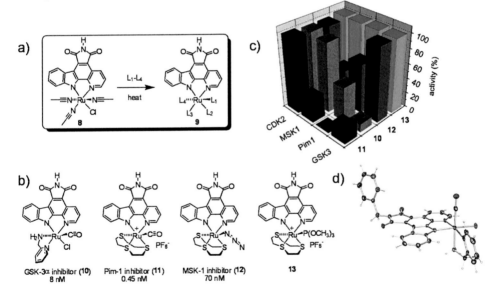

Figure 8. (a) Synthesis of ruthenium pyridocarbazole complexes **9** by ligand substitution chemistry from the common precursor **8**. (b) Selected complexes derived from **8**. IC_{50} values were measured at 100 μM ATP. (c) Inhibitory activities of the organoruthenium compounds **10-13** at a concentration of 100 nM against the protein kinases GSK-3α, Pim-1, MSK1, and CDK2/CyclinA. ATP concentration was 100 μM. (d) Crystal structure of the N-benzylated derivative of **10**. ORTEP drawing with 40% probability thermal elipsoids.

Conclusion

With GSK-3 and Pim-1 as model kinases we have established that organometallic scaffolds are highly promising scaffolds for the design of kinase inhibitors. Accordingly, we have identified low nanomolar and even picomolar inhibitors for these kinases, demonstrated their exceptional kinase selectivities, have verified their usefulness in mammalian cell culture and model organisms such as zebrafish and frog embryos, and we confirmed their binding modes with cocrystal structures of ruthenium inhibitors bound to the ATP-binding site of Pim-1. Furthermore, we have developed tools that allow to rapidly synthesize a diverse selection of organometallic compounds and to screen them for their kinase activities. The discovered organometallic GSK-3 and Pim-1 inhibitors are some of the most potent and selective inhibitors for kinases know to date and therefore these results validate the strategy of exploring organometallic scaffolds for the design of enzyme inhibitors. Future work has to focus on octahedral metal complexes and expand the investigation to other biological targets.

References

[1]   Alan, H., Lipkus, A.H., Yuan, Q., Lucas, K.A., Funk, S.A., Bartelt III, W.F., Schenck, R.J., Trippe, A.J. (2008) Structural Diversity of Organic Chemistry. A Scaffold Analysis of the CAS Registry. *J. Org. Chem.* **73**:4443 – 4451.

[2]   Bemis, G.W., Murcko, M.A. (1996) The properties of known drugs. 1. Molecular frameworks. *J. Med. Chem.* **39**:2887 – 2893.

[3]   Meggers, E. (2007) Exploring biologically relevant chemical space with metal complexes. *Curr. Opin. Chem. Biol.* **11**:287 – 292

[4]   Dyson, P.J., Sava, G. (2006) Metal-based antitumour drugs in the post genomic era. *Dalton Trans.* **16**:1929 – 1933.

[5]   Hartinger, C.G., Zorbas-Seifried, S., Jakupec, M.A., Kynast, B., Zorbas, H., Keppler, B.K. (2006) From bench to bedside – preclinical and early clinical development of the anticancer agent indazolium trans-[tetrachlorobis(1 H-indazole)ruthenate(III)] (KP1019 or FFC 14A). *J. Inorg. Biochem.* **100**:891 – 904.

[6]   Meggers, E., Atilla-Gokcumen, G.E., Bregman, H., Maksimoska, J., Mulcahy, S.P., Pagano, N., Williams, D.S. (2007) Exploring Chemical Space with Organometallics: Ruthenium Complexes as Protein Kinase Inhibitors. *Synlett.* **8**:1177 – 1189.

[7]   Huwe, A., Mazitschek, R., Giannis, A. (2003) Small molecules as inhibitors of cyclin-dependent kinases. *Angew. Chem. Int. Ed.* **42**:2122 – 2138.

[8]   Bregman, H., Williams, D.S., Atilla, G.E., Carroll, P.J., Meggers, E. (2004) An Organometallic Inhibitor for Glycogen Synthase Kinase 3. *J. Am. Chem. Soc.* **126**:13594 – 13595.

[9]   Debreczeni, J.É., Bullock, A.N., Atilla, G.E., Williams, D.S., Bregman, H., Knapp, S., Meggers, E. (2006) Ruthenium Half-Sandwich Complexes Bound to Protein Kinase Pim-1. *Angewandte Chem. Int. Ed.* **45**:1580 – 1585.

[10]   Atilla-Gokcumen, G.E., Williams, D.S., Bregman, H., Pagano, N., Meggers, E. (2006) Organometallic Compounds with Biological Activity: A Very Selective and Highly Potent Cellular Inhibitor for Glycogen Synthase Kinase 3. *ChemBioChem* **7**:1443 – 1450.

[11]   Maksimoska, J., Williams, D.S., Atilla-Gokcumen, G.E., Smalley, K.S.M., Carroll, P.J., Webster, R.D., Filippakopoulos, P., Knapp, S., Herlyn, M., Meggers, E. (2008) Isostructural Ruthenium and Osmium Complexes Display Highly Similar Bioactivities. *Chem. Eur. J.* **14**:4816 – 4822.

[12]   Xie, P., Williams, D.S., Atilla-Gokcumen, G.E., Milk, L., Xiao, M., Smalley, K.S.M., Herlyn, M. Meggers, E. Marmorstein, R. (2008) Structure-Based Design of an Organoruthenium Phosphatidyl-inositol-3-kinase Inhibitor Reveals a Switch Governing Lipid Kinase Potency and Selectivity. *ACS Chemical Biology* **3**:305 – 316.

[13]   Williams, D.S., Atilla, G.E. Bregman, H., Arzoumanian, A., Klein, P.S., Meggers, E. (2005) Switching on a signaling pathway with an organoruthenium complex. *Angew. Chem. Int. Ed.* **44**:1984 – 1987.

[14]   Smalley, K.S.M., Contractor, R., Haass, N.K., Kulp, A.N., Atilla-Gokcumen, G.E., Williams, D.S., Bregman, H., Flaherty, K.T., Soengas, M.S., Meggers, E., Herlyn, M. (2007) An Organometallic Protein Kinase Inhibitor Pharmacologically Activates p53 and Induces Apoptosis in Human Melanoma Cells. *Cancer Res.* **67**:209 – 217.

[15]   Bregman, H., Carroll, P.J., Meggers, E. (2006) Rapid Access to Unexplored Chemical Space by Ligand Scanning around a Ruthenium Center: Discovery of Potent and Selective Protein Kinase Inhibitors. *J. Am. Chem. Soc.* **128**:877 – 884.

 **Beilstein-Institut**

Systems Chemistry, May 26th – 30th, 2008, Bozen, Italy

MODELLING FOR REGENERATIVE MEDICINE: SYSTEMS BIOLOGY MEETS SYSTEMS CHEMISTRY

DAVID A. WINKLER[*], JULIANNE D. HALLEY, FRANK R. BURDEN

CSIRO Molecular and Health Technologies,
Private Bag 10, Clayton 3168, Australia

E-Mail: *dave.winkler@csiro.au

Received: 5th September 2008 / Published: 16th March 2009

INTRODUCTION

Complex systems science is making substantial contributions to the study of biological systems, and has made a substantial contribution to the new field of systems biology. Systems biology focuses on the systematic study of complex interactions in biological systems using an integrative rather than reductionist perspective. One of the goals of systems biology is to study, model, and understand new emergent properties of biological systems from a complex systems perspective [1, 2]. This integrative approach to biology is generating substantial benefits in facilitating study of larger more complicated systems, providing improved understanding of nonlinear system properties, and provides an ability to model systems at appropriate levels of detail where the model is matched to data density and research questions. Various aspects of systems biology have been reviewed recently [3 – 11]. Chemistry has lagged behind most other disciplines in adopting complex systems approaches, possibly because it has largely been a reductionist science, and reductionist approaches have been very successful. Adopting a complementary complex systems approach to chemistry will build on this success to study more complex matter.

Paradigm Shifts

Chemistry focuses on the synthesis and properties of relatively small sets of molecular species but is now increasingly embracing the generation and study of larger, more diverse systems of molecules [12]. Chemistry is becoming increasingly multidisciplinary, embracing important new research fields where chemistry, physics, biology and materials science overlap. Emerging research fields such as nanomaterials and self-assembly will benefit from an integrative, complex systems approach that is complementary to the reductionist methods that characterize the discipline of chemistry [13]. Reductionist and deterministic concepts like proteins and ligands as rigid 'lock and key' systems, cells as molecular machines, and proteins as relatively rigid structures are increasingly at odds with new spectroscopic data, computational experiments, and high throughput experimental results. An adaptive, dynamic description of small molecules (chemistry), and larger molecules (biology but now increasingly chemistry) is required. As Kurakin elegantly states [14], there is a global crisis of the mechanistic, deterministic paradigm in life sciences that a complex systems description of molecular processes can resolve. In his classic work on the nature of scientific revolutions, Kuhn states that the success of a dominant paradigm brings about its own crisis and necessitates a paradigm shift, as the advances in technology and methods lead to a widespread accumulation of experimental data that cannot be readily explained and accommodated within the existing conceptual framework [15]. Kurakin's papers give some diverse and important examples of these [14, 16 – 20].

This paper is not intended to be an exhaustive review of complex systems and their application to chemistry, rather a summary of the main concepts, important areas of application, and examples in an important area of overlap between chemistry and biology. The first section reviews the major elements of complex systems science from a chemistry perspective, summarizes the relatively small research effort in chemical complex systems, and indicates where a complex systems approach is likely to provide substantial benefit to new and existing chemical and biological challenges. By analogy with systems biology, we and others define the integrative study of very complicated chemical systems using the tools and concepts of complex systems science as *systems chemistry*, a term first employed by von Kiedrowski to define the chemical origins of biological organization [21]. Ludlow and Otto published the first review of systems chemistry as a subject in 2008 [22], although Whitesides noted the need to apply complexity concepts to chemistry in 1999 [12]. The second section of the paper focuses on the importance of scale in efficient modelling of complex systems, and discusses how Bayesian methods can be used to choose an appropriate level of scale for models. One of the most exciting new areas of chemical and biological research is the understanding and molecular control of stem cell fate for use in regenerative medicine. Complex systems tools and concepts are providing a novel, complementary approach to understanding and controlling stem cell fate. The third part of the paper illustrates how this complexity can be applied to modelling processes driving stem cell fate decisions.

COMPLEX SYSTEMS

Complex systems science is a rapidly growing new paradigm for understanding and modelling extremely complicated systems (physical, social, economic, biological etc), and for discovering common mechanisms in apparently diverse phenomena. They generally exhibit nonlinear dynamic behaviour caused by many interactions among the system components, and have so-called emergent properties that are difficult or impractical to understand or predict from knowledge of the components from which the system is constructed. Some complex systems can evolve in ways that are very sensitive to initial conditions. This behaviour can be described by nonlinear differential equations. Examples of complex systems are found in virtually all disciplines of science – from molecular self-organization and self-assembly to weather patterns, social systems, ecosystems, electricity grids, economies, and more esoteric properties like self-awareness, emotion, and life itself. Complex systems have been reviewed extensively in the recent literature [23 – 39]. Although there is still some debate about properties exhibited by complex systems, most exhibit a number of important properties that distinguish them from simple systems.

Criticality and nonlinearity

Complex systems often undergo an abrupt change in properties when a threshold value of some system parameter is crossed. This event is analogous to a phase transition that results from component interactions within the system switching from local connectivity to global connectivity. Many complex systems exhibit this type of behaviour, which is characterized by order and stability below the critical point and instability and chaos above the critical point. Phase transitions, crystallization, and alignment of magnetic domains in materials are common examples of systems exhibiting critical behaviour. Cramer and Booksh published one of the few reviews on chaos theory applied to chemistry, and list several interesting classes of chemical systems in which chaotic behaviour occurs [40].

Autocatalytic systems, which abound in chemistry and biology, are examples of nonlinear systems. The most familiar example of this type of behaviour is the complex spiral pattern observed in the BZ (Belousov-Zhabotinsky) reaction (Fig. 1). Although often observed as temporally varying series of patterns, reactions of this type also generate spatial patterns. Reaction-diffusion processes have been proposed as a mechanism for pattern formation in animals (e. g. tiger stripes), as a mechanism that generates body plans during embryogenesis.

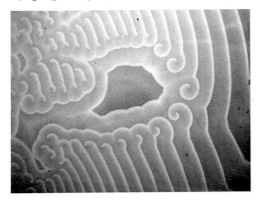

Figure 1. Spiral waves in BZ reaction

Self-organization and self-assembly

Interactions between large numbers of components may generate self-organized and self-assembled structures. Unfortunately, these terms are used in different ways in different scientific disciplines, and are even used interchangeably. A simple way of distinguishing them is to define self-assembly as a process that generates a structure that is in static equilibrium and is thermodynamically more stable than its components, and the assembly is driven by this energy difference [41]. Self-organizing systems increase their internal order over time implying a decrease in system entropy, so that self-organizing systems cannot be closed. The second law of thermodynamics requires energy transfers across the boundary and there is a corresponding increase in the entropy of the environment in which the system is embedded [33]. The self-organized structure is maintained in dynamic equilibrium and decays if the energy source is removed. A recent paper discusses the relationship between self-organization and self-assembly and gives examples of how these definitions apply across multiple disciplines [41]. Examples of systems that are difficult to categorize and that cross boundaries are also given. Understanding the mechanisms that underlie self-assembly and self-organization will enhance our ability to design materials and new technology for important fields such as nanotechnology, tissue engineering, and biomaterials. Complex systems concepts and modelling tools provide additional methods for understanding the behaviour of these systems.

Emergence and scale

Reductionist methods typically attempt to explain how a system behaves by considering the properties of the system components and their interactions. Complex systems approaches highlight emergent properties of a system and allow very complicated systems to be viewed at multiple levels of scale. Appropriate scales capture the emergent macroscopic properties of interest, without requiring a full understanding of mechanisms at finer levels of detail. In most cases, an ultimate fine-grained model of a complicated system is not possible because of the difficulty of modelling at this level of scale and the large amount of information required to build and validate such models. Generally, time, cost, or other resource limitations prevent this information becoming fully available. Complex systems often exhibit properties that are self-similar or scale-invariant (fractal) [42–45]. Power law behaviour in which a log-log plot of frequency versus magnitude of the property is linear, is observed in a diverse range of physical and biological complex systems [36, 46, 47]. The slopes of the power law graphs (the exponent of the power law) for different systems tend to cluster, generating *Universality Classes* that may have common underlying mechanisms [48–50]. Clearly, emergence, scale and complexity are interrelated and, as they are difficult concepts to define precisely, there is considerable latitude in how the concepts are employed across multiple disciplines. Recent work has attempted to generate more precise, widely applicable definitions for terms such as emergence, scale, self-assembly and self-organization [13, 41, 51, 52].

Emergent, holistic properties of complex systems are usually the ones that we observe and attempt to engineer. As they are generated by the relatively simple nonlinear interactions of many system components, these emergent properties are to some extent unexpected or surprising. Examples of emergent properties include the self-organized convection cells in heated liquids (Fig. 2), the oscillation of species in predator-prey systems, and weather patterns. Emergence and self-organization are intimately related, and recent work has described how emergence may be characterized in terms of self-organization [52].

Figure 2. Hexagonal, self-organized convection cells (Bénard cells) in a heated fluid.

Emergent properties can vary in their 'depth' [41, 52]. Simple emergent properties include gas pressure resulting from the interaction between atoms or molecules and their environment. More deeply emergent chemical properties include the formation of entirely new molecular species with novel properties when molecules participate in chemical reactions. Physical properties of molecules (e. g. aromaticity) can also be considered emergent, as it is very difficult or impossible to predict them from first principles given the properties of their components (atoms) and the interactions in which these components participate. However, emergent properties can often be predicted at a different level of scale (a coarser level of description). For example, the pressure of a gas can be calculated using a dynamical simulation that accounts for the speed and direction of many atoms or molecules, but is more readily calculated at a coarser level of scale using the gas law [53].

Networks and interactions

Networks are very important paradigms for understanding interactions in complex systems as they are a compact way of describing interactions between components. They have been extensively studied in many fields other than chemistry. If a complex chemical system consists of components that interact in a directed or non-directed way, this can be easily presented as a network or graph. Although graph theory has been long been applied to chemical systems such as molecules, relatively little work has been done on how network representations of molecules and chemical reactions can yield new insight into the properties of molecules. In systems biology, network paradigms are widely used to understand genetic regulation, signalling, and protein-protein interactions. There has been recognition recently that an understanding of network properties is necessary to properly exploit biological vulnerabilities and target drugs more effectively. Network concepts are therefore becoming increasingly important in medicinal chemistry and their applications to drug design have been reviewed recently [54 – 58].

SYSTEMS CHEMISTRY

Compared to systems biology, the application of complex systems science to chemistry is almost non-existent. This is surprising given that Ilya Prigogine was awarded the Nobel Prize in Chemistry in 1977 for work on nonlinear dynamical systems in chemistry that helped to establish the foundations of complex systems science. Several other Nobel Prizes have been awarded for research relating to self-assembling systems (Crick, Watson and Wilkins, Medicine 1962; Krug, Chemistry 1982, Lehn, Chemistry, 1987) that build on complex systems frameworks. Clearly, Watson, Crick and Wilkins' work on understanding the structure and self-assembly of DNA led to a paradigm shift in our understanding and revolutionized the fields of biology and medicine. However, their work had a relatively smaller impact on chemistry, a discipline that is well equipped to describe the interactions that control DNA's structure. A few other chemists, notably Testa, Kier and Belousov and Zaboutinsky, explored complex behaviour in chemistry but the area did not develop in a sustainable way and declined in prominence. In a sense, the field was ahead of its time, and did not offer solutions to problems that reductionism could not tackle adequately. Within the past decade, interest in applying complexity to chemistry has increased, and several workers, notably Whitesides, have written seminal papers. Whitesides proposed that almost every-thing of interest in chemistry is complex, by the above definitions of complexity [12]. For example a single step in a multistep synthesis involves $\sim 10^{22}$ molecules of several types, a myriad of interacting nuclei and electrons and $\sim 10^{24}$ molecules of solvent. As such real systems cannot be dealt with at the molecule level, synthetic chemistry has employed approximations, rules, and heuristics. Complex systems paradigms can provide additional, complementary methods for understanding and dealing with such immense complexity.

CONTEMPORARY CHEMICAL RESEARCH WHERE A COMPLEX SYSTEMS APPROACH MAY BE RELEVANT

Most of the interesting new areas of chemistry are discipline spanning, and will benefit from a complementary complex systems approach.

Nonlinear dynamical systems and pattern formation

Nonlinear dynamical systems can exhibit chaotic behaviour, often modelled using differential equations akin to those describing chemical kinetics. Spatial and cellular automata methods are being applied increasingly to dynamical systems. As discussed, some chemical reactions can behave in this manner. Although the most recognised example is the BZ reaction [59] (see Figure 2), many other dynamic, pattern-forming reactions such as the Briggs-Raucher reaction are known. Recently, reaction-diffusion or oscillating reactions have been used to generate novel materials such as a self-walking gel [60, 61].

Fractal materials

Systems that exhibit fractal or self-similar behaviour are found widely in nature – snowflakes, lightning, crystals, river networks, ferns, blood vessels and coastlines. Real world objects show fractal behaviour over an extended, but finite, scale range [108]. Fractal behaviour is common in nano-structured materials, where self-similarity provides unique properties such as giant enhancement of nonlinear optical properties, and related effects that underlie surface-enhanced Raman scattering [62]. Fractal effects are also important in electrochemistry [63].

Supramolecular polymers

Conventional polymers are linked by permanent covalent bonds. Supramolecular polymers are a new class of polymers in which the linkages are reversible. The dynamic properties of these materials open up the prospect of many new applications. The most important step in developing practical supramolecular polymers was the discovery of the 2-ureido-4[1 H]-pyrimidinone unit that could form four hydrogen bonds. This monomer has self-complementary allowing it to self-assemble into polymers with novel properties, such as the ability to self-heal when cut [64].

Self-replicating systems

Self-replicating systems clearly exist in biology and there has been much speculation on the types of autocatalytic systems that may have arisen on the primordial earth that may have led to self-replicating molecules and life. Kauffman argued that a sufficiently large pool of chemicals may be intrinsically capable of generating autocatalytic cycles and self-replicating molecules [65]. Seminal work by Wintner *et al.* showed how simple organic structures can catalyse their own formation. Self-complementarity is the key to this autocatalytic behaviour [66]. Design of self-replicating molecules is very difficult and progress in the field has been slow. The recent review by Paul and Joyce summarizes progress in the area [67].

Artificial life

Artificial life and the closely related field of synthetic biology covers a range of research areas, many of them intimately involved with chemistry. It spans synthesis of unnatural organic molecules that function in living systems, through biomimetic chemistry that produces synthetic molecules that recapitulate the behaviour of natural analogues, to the greatest challenge of recreating synthetic chemical systems that exhibit inheritance, genetics and evolution. Recent reviews and papers by Benner, Rosen, and Cornish-Bowden provide a summary of progress in artificial life [68 – 70].

Dynamic combinatorial libraries

Dynamic combinatorial chemistry is defined as combinatorial chemistry under thermodynamic control, in which all constituents are in equilibrium. The library members are therefore constantly interconverting via reversible chemical reactions. The composition of the library can therefore be altered by changing factors in the environment [71]. For example, dynamic libraries have been exploited in drug discovery. When members of the library interact favourably with a biological target, the composition of the library shifts and results in enrichment of compounds that bind to the target protein. The environment can therefore 'select' or enrich the library with compounds that are more 'fit'. Given the enormous size of drug-like chemistry space, dynamic combinatorial libraries offer a more efficient method of exploring this space to identify novel drug leads than covalently bonded combinatorial libraries. The technique is currently limited by the small number of reversible reactions that are biocompatible.

Evolutionary methods

Genetic algorithms and genetic programming are also evolutionary, adaptive tools that are finding increased applications in chemistry and biology. Genetic algorithms are very efficient methods for exploring vast search spaces such as those involved in protein folding, or drug-like chemical space. When applied to chemistry, they involve a string representation of a molecule or molecular property (e.g. a SMILES string, torsion angle list etc), a mutation operator (e.g. swap one atom type or functional group for another, add or delete an atom or functional group, change a torsion angle etc), and a fitness function (e.g. biological activity, desired protein fold etc). The algorithm generates families of mutants that are assessed by the fitness function. The fittest are retained and less fit discarded. The cycle repeats starting with the fittest individuals from the previous cycle, and stops when some criterion is matched. They have been used to find optimum sets of molecular descriptors for modelling of drug activity, for searching chemical databases for leads [73 – 75], and for optimising combinatorial libraries [76], discovery and optimisation of new catalytic materials [77 – 79], design of peptide mimetopes [78, 79], multiobjective optimisation [80, 81], synthesis planning [82 – 85], and in many other chemical applications.

Drug targeting

Interactions between molecules (e. g., drugs and proteins, proteins and DNA, etc.) in living systems can be described by networks. Very recently, network approaches have been applied to understand why the efficacy of new drug discovery is declining [54, 55, 86]. It has long been recognized that almost all drugs hit multiple targets, which is problematic as it leads to off-target effects and toxicity. Drugs are often designed to hit a specific target that is presumed to induce the desired phenotypic change, but it has only recently been appreciated that the robustness of biological networks resists such perturbations and may allow systems to adapt to minimize the impact. If we understand the robustness of biological networks, we could better engineer drugs to hit multiple intentional targets to affect the best outcome.

Supramolecular, self-organizing and self-assembling systems

Self-assembling, supramolecular, and self-organizing materials are currently of considerable interest to chemists [87 – 90]. New research fields of nanoscience and smart materials aim to design materials and molecular devices that can be assembled from the 'bottom up' rather than 'top down'. Chemical or biological materials have been shown to undergo self-assembly, although a complete understanding of assembly mechanisms is still lacking, making design challenging. Self-assembly has been applied to electronic device manufacture [91], viral capsid modelling [92], nanoparticle assembly [93 – 95], peptides self-assembly, nanotubes, glycolipids, carbohydrates, polymers, crystals [94, 96 – 101], and many other materials[30]. Aberrant self-assembly provides the mechanism for many protein folding diseases such as prion and Alzheimer's diseases, amyloidoses, cataracts, and diabetes [102, 103]. New complexity-based tools, such as agent-based models, have an increasing important role to play in design of self-assembling materials [13, 104].

Self-organization also applies to chemistry but is less prominent in the chemical literature largely due to the tendency of many authors to describe self-assembly as self-organization[39]. Most examples of true self-organization involve biological systems because they are invariably open [33, 105, 106]. A chemical example (although biology-based) is the dynamic instability of microtubules [107]. Formation of microtubules involves a spontaneous self-assembly of tubulin, but the dynamics of the process also involves disassembly using energy librated when tubulin hydrolyses GTP (tubulin is an enzyme as well as structural protein) [108]. An example of a self-organized tubulin structure is given in Figure 3.

Cram and Lehn conducted seminal work on self-assembly, establishing the field of supra-molecular chemistry involving the generation of complex structures from reversible non-covalent bonds These structures exploit hydrogen bonding, metal coordination, hydrophobic forces, van der Waals forces, π-π interactions and electrostatic effects [109].

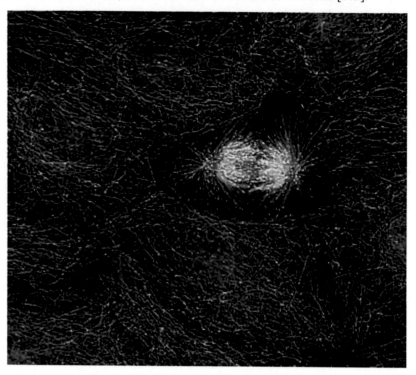

Figure 3. A self-organized self-assembly of microtubules.

MESCOSCALE MODELLING – CHOOSING THE CORRECT SCALE

Some complex systems appear to be 'computationally irreducible', raising the question of how we model emergent properties of such systems. Recent work on diverse complex systems suggests sparse, mesoscale, or 'coarse grained' models, that omit most of the detail and concentrate only on the essential pattern forming process, can provide a path forward. Choosing too low a level of detail (too coarse) results in models that have a limited ability to generalize, and can predict only broad features of the system. Choosing a too detailed level of scale produces more complicated models but requires much more data. As many modelling exercises are data limited, overly complex models pose problems for parameterisation, validation, and they are prone to overfitting.

Biologically, coarse graining of sensory information is essential so that organisms can function without being overloaded. For example, in the visual system the eye does substantial feature selection and classification before information is sent to the brain. If mesoscale descriptions can adequately model the essential (emergent) properties of complex systems, how do we find the correct level of scale that matches the usefulness of the predictions to the availability of data? We adopt two approaches, one specific for biology and one more generally applicable.

Sparse feature selection and modelling

Generically, mesoscale models can be generated using sparse feature selection methods that select in a context-dependent manner, the most relevant aspects of the system under study. Prominent of these are Bayesian methods like expectation maximization algorithms, described elegantly by Figeuiredo [110].

Neural networks are another adaptive, learning method that is finding increased application in chemistry and biology [58, 111 – 116]. Neural networks are algorithms that learn patterns or associations between mathematical representations of objects and emergent properties of systems that are comprised of, or interact with these objects. An important area where neural networks have contributed to complexity-based modelling of chemical systems, is in QSAR and chemometrics. Neural networks have been used widely in chemistry and pharmacology for finding relationships between molecules and their biological or physicochemical properties. They have an inherent ability to model a complex system at a sufficiently coarse grained scale that matches the model complexity to the availability of data. This is particularly important in medicinal chemistry, which is often data-poor [12]. Neural networks are platform technologies that have been used to model a myriad of emergent properties of chemical and biological complex systems, such as drug target activity [117 – 121], drug-like (ADME) properties [122 – 126], immune modulation [127, 128], physicochemical properties [129 – 136], toxicity [112, 137 – 142].

Mesoscale models of regulatory interactions

Although there is vigorous debate [143] about the architecture of biological networks, many are scale-free due to the intrinsic robustness of such architectures. Mesoscale models of cell regulatory networks can be derived by focusing on key network hubs and genes controlled by these hubs that are relevant to the biological process under study. Mesoscale regulator models can represent many types of interacting components in biology. In gene regulation, the hubs represent the set of genes most relevant to a particular biological process such as differentiation, apoptosis or self-renewal. Mesoscale regulatory networks can be deduced by a 'bottom up' reductionist analysis of gene microarray data, or data from other experiments such as chromatin immunoprecipitation (ChIP). An alternative 'top down' approach is to generate 'blank' mesoscale networks that are subsequently trained to recapitulate experimental data or predict the outcomes of new experiments. The latter approach was first described by Geard and Wiles, and is a substantially novel analogue applied to modelling of quantitative gene regulation in *C. elegans* embryogenesis [144, 145]. The approach is illustrated in Figure 4.

This coarse-grained approach to cell differentiation allows us to rise above the details of biochemical experiments to produce a map or 'fate matrix' of cell behaviour that reflects the integration of multiple layers of complexity in cellular processes. Given a set of target genes that characterize each cell embryogenesis pathway, a mesoscale regulatory network model can be trained using gene microarray data, so that its output mimics changes in gene activity as cells divide and/or differentiation. Once trained, the network architecture and weight matrix represent a coarse-grained model of the regulatory network controlling cell fate. Each gene in the model interacts with the weight matrix, reflecting the fact that the transcription of a gene is the result of integrating the cell's biochemical state rather than the action of any single gene product. One set of master genes (i.e. one expression state) regulates the transcription of the cell's genes, leading to a new state. We aim to annotate these mesoscale models with experiments identifying regulatory interactions, to produce a model of sufficient detail to understand and control cell fate [146].

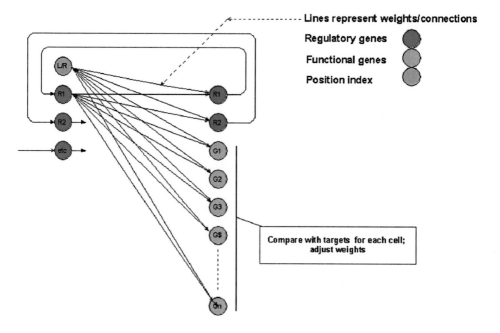

Figure 4. Recursive neural network structure, showing how the regulatory genes and relative position index control the expression levels of genes in the various cells in the lineage.

MODELLING STEM CELL FATE DECISIONS

Complex systems science coupled with interdisciplinary approaches provide scope to integrate the many different layers of complexity that characterise complex biological systems. Hence, important areas of research such as stem cell biology a growing literature views stem cell function within the framework of a complex adaptive network, rather than the traditional deterministic and reductionist approach that focuses on single stem cells [147]. We summarize several areas where sparse methods have been applied to build mesoscale models of properties of stem cell or related multipotent cells.

Stemness genes in human embryonic stem cells (hESCs)

Laslett *et al.* isolated sub-populations of embryonic stem cells that differed slightly in their degree of 'stemness' (pluripotency) [148]. The experiments aimed to identify genes that maintained pluripotency in hESC and those that drive early commitment to other lineages. Flow cytometry was used to select a small set of pluripotent hESCs using fluorescent antibody markers whose expression correlated with the pluripotency marker Oct4. This very stemlike subpopulation was further sorted into hESC populations at four, graded, very early stages of differentiation (P4-P7) (Fig. 5). Gene expression microarrays were generated from each of the populations. Sparse Bayesian feature selection algorithms identified a small number of markers that correlate with the pluripotency of the hESCs, and were classifier genes with a putative key role in the maintenance of pluripotence. They could achieve this classification with high statistical significance, suggesting the genes played a prominent role in stemness.

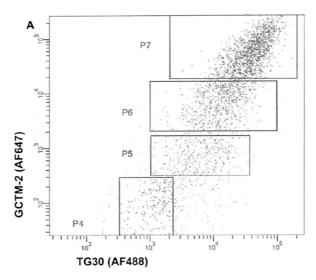

Figure 5. Flow cytometric experiments identifying the four classes of hESCs.

Erythroid differentiation

Welch and co-workers recently reported experiments on estradiol-induced differentiation along the erythroid pathway over 30 hours [149]. This trajectory starts from the late burst-forming unit-erythroid stage and progresses to the orthochromatic erythroblast stage. Estradiol was used to trigger the expression of the gene GATA-1, initiating synchronous differentiation in the population. Microarrays were generated for populations of cells at each time point during the experiment. Five classifier genes were found that correlate with the stage of differentiation down the lineage. These genes were highly reliable markers for the differentiation process, and are likely to play an important role in regulation of the erythroid pathway.

Mesoscale network modelling of mouse embryogenesis

We modelled murine embryogenesis from the zygote to the blastocyst stage (Fig. 6) using a 'top down' mesoscale regulatory modelling framework.

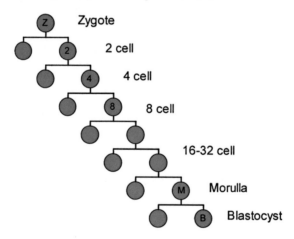

Figure 6. The lineage diagram for mouse embryogenesis. Cell labels are shown.

Hamatami *et al.* measured the expression level of genes in carefully prepared samples of cells from the unfertilized egg to the blastocyst using gene microarrays [150] (see Figure 6). We modelled 135 genes from the mid-preimplantation gene activation (MGA) cluster for each cell type in the embryogenesis pathway. We first trained the model using the gene expression targets for each of the 135 genes in the cell types at the seven time points. We then assessed the ability of the model to predict gene expression in cells in the pathway. The mesoscale regulatory network model recapitulated quantitative gene expression profiles within the cell lineage surprisingly high fidelity. We also generated models that predicted expression in cells at an intermediate point in the lineage (interpolation) and at the end of the lineage (extrapolation) when these were not used in training.

Figure 7 summarizes these predictions. Figure 7A shows how well the model is able to recapitulate the quantitative expression levels of all genes in all cells in the training set. Figure 7B shows the most stringent test where the expression levels of the last three cell divisions in the trajectory are predicted from a model trained on the first four cells in the lineage. While the error is substantially higher than in the single cell expression prediction, the predictions nevertheless have high statistical significance.

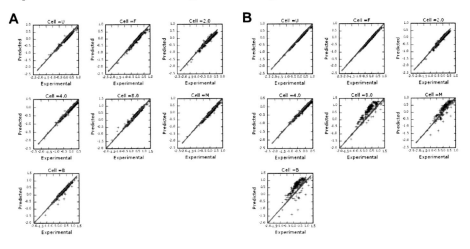

Figure 7. Predicted versus experimental log2 fold expression levels for genes in the seven cell types. (**A**) Training set. (**B**) prediction of 8-cell, morula and blastocyst expression for model trained on first 4 cells in embryogenic pathway.

Why sparse models work

Our sparse feature selection methods and mesoscale regulatory models support the common observation that a relatively small number of factors control the fate of stem cells – as few as three to six. Hart *et al.* also illustrated a similar sparse transcription factor requirement when modelling the yeast cell cycle transcription network [151]. They initially used 204 transcription factors to model the temporal variation of genes across the cell cycle. They pruned their network models progressively, removing the least relevant transcription factors in turn, and monitored the performance of the model. They found a gradual and limited degradation in model performance until five transcription factors remained, after which the performance of the model collapsed (Fig. 8).

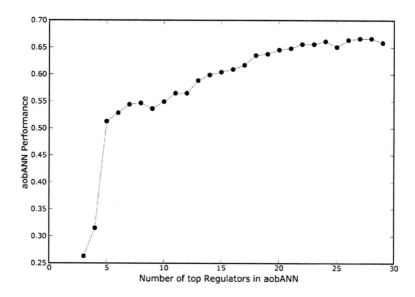

Figure 8. Performance of neural network model of temporal variation in gene expression during cell cycle. Note the dramatic drop in performance when the number of transcription factors is 5 or less.

CONCLUSIONS AND THE FUTURE

The evolution of chemistry away from manipulation of single molecules towards description and manipulation of complex systems of molecules is one of the key driving forces for embracing systems chemistry. In this way, the move to the study of more complex matter may counter intuitively generate new types of problems that are sufficiently simple when viewed in a complexity framework that they can be tackled in an analytical sense [12]. We have arrived at Kuhn's crisis of the dominant paradigm in chemistry, a period that is always followed by an exciting period of extraordinary research and new discoveries [14]. However, we can only achieve this by recognizing the opportunity and acting on it. Ludlow and Otto's call for chemists to embrace complexity and recognize systems chemistry is a very timely challenge to expand this important new paradigm for understanding chemical problems [22].

Chemists may finally capitalize on the powerful but in some respects premature possibilities that the Nobel Laureates presented to us, and move more rapidly into the exciting worlds of new chemistries, materials, and biology.

REFERENCES

[1]   Kitano, H. (2002) Computational systems biology. *Nature* **420**:206 – 10.

[2]   Kitano, H. (2002) Systems biology: a brief overview. *Science* **295**:1662 – 4.

[3]   Way, J.C., Silver, P.A. (2007) Systems engineering without an engineer: Why we need systems biology. *Complexity* **13**:22 – 29.

[4]   Fisher, J., Henzinger, T.A. (2007) Executable cell biology. *Nat. Biotechnol.* **25**:1239 – 49.

[5]   Aigner, T., Haag, J., Zimmer, R. (2007) Functional genomics, evo-devo and systems biology: a chance to overcome complexity? *Curr. Opin. Rheumatol.* **19**:463 – 70.

[6]   Marguet, P., Balagadde, F., Tan, C.M., You, L.C. (2007) Biology by design: reduction and synthesis of cellular components and behaviour. *J. Roy. Soc. Interface.* **4**:607 – 23.

[7]   Quackenbush, J. (2007) Extracting biology from high-dimensional biological data. *J. Exp. Biol.* **210**:1507 – 17.

[8]   Materi, W., Wishart, D.S. (2007) Computational systems biology in drug discovery and development: methods and applications. *Drug Disc. Today.* **12**:295 – 303.

[9]   van Regenmortel, M.H.V. (2007) The rational design of biological complexity: A deceptive metaphor. *Proteomics* **7**:965 – 75.

[10]   Lin, J., Qian, J. (2007) Systems biology approach to integrative comparative genomics. *Expert. Rev. Proteomics* **4**:107 – 19.

[11]   Palumbo, M.C., Farina, L., Colosimo, A., Tun, K., Dhar, P.K., Giuliani, A. (2006) Networks everywhere? Some general implications of an emergent metaphor. *Curr. Bioinfo.* **1**:219 – 34.

[12]   Whitesides, G.M., Ismagilov, R.F. (1999) Complexity in chemistry. *Science* **284**:89 – 92.

[13]   Lehn, J.-M. (2002) Toward self-organization and complex matter. *Science* **295**:2400 – 3.

[14]   Kurakin, A. (2007) Self-organization versus Watchmaker: ambiguity of molecular recognition and design charts of cellular circuitry. *J Mol Recog.* **20**:205 – 14.

[15]   Kuhn, T. (1996) The Structure of Scientific Revolutions. Chicago: University of Chicago Press.

[16]   Kurakin, A. (2004) Self-organization versus watchmaker: stochasticity and determinism in molecular and cell biology. Novato lectures.

[17] Kurakin, A. (2005) Self-organization vs. Watchmaker: stochastic gene expression and cell differentiation. *Dev. Genes Evol.* **215**:46 – 52.

[18] Kurakin, A. (2005) Stochastic cell. *IUBMB Life* **57**:59 – 63.

[19] Kurakin, A. (2005) Self-organization versus watchmaker: stochastic dynamics of cellular organization. *Biol. Chem.* **386**:247 – 54.

[20] Kurakin, A. (2006) Self-organization versus Watchmaker: Molecular motors and protein translocation. *Biosys.* **84**(1):15 – 23.

[21] Stankiewicz, J., Eckardt, L.H. (2006) Chembiogenesis 2005 and Systems Chemistry Workshop. *Angew. Chem. Int. Ed.* **45**:342 – 4.

[22] Ludlow, F.R., Otto, S. (2008) Systems Chemistry. *Chem. Soc. Rev.* **37**:101 – 8.

[23] Batty, M. (2008) The size, scale, and shape of cities. *Science* **319**:769 – 71.

[24] Luzzi, R, Vasconcellos, A.R., Ramos, J.G. (2007) Non-equilibrium statistical mechanics of complex systems: An overview. *Rivista Del Nuovo Cimento* **30**:95 – 157.

[25] Mara, A., Holley, S.A. (2007) Oscillators and the emergence of tissue organization during zebrafish somitogenesis. *Trends Cell Biol.* **17**:593 – 9.

[26] Weber, B.H. (2007) Emergence of life. *Zygon.* **42**:837 – 56.

[27] Foote, R. (2007) Mathematics and complex systems. *Science* **318**:410 – 2.

[28] Eckmann, J.P., Feinerman, O., Gruendlinger, L., Moses, E., Soriano, J., Tiusty, T. (2007) The physics of living neural networks. *Physics Rep-Rev Section Phys Lett.* **449**:54 – 76.

[29] Rammel, C., Stagl, S., Wilfing, H. (2007) Managing complex adaptive systems – A co-evolutionary perspective on natural resource management. *Ecolog. Econom.* **63**(1):9 – 21.

[30] Balazs, A.C. (2007) Modelling self-assembly and phase behaviour in complex mixtures. *Ann. Rev. Phys. Chem.* **58**:211 – 33.

[31] Almaas, E. (2007) Biological impacts and context of network theory. *J. Exp. Biol.* **210**:1548 – 58.

[32] Mitchell, M. (2006) Complex systems: Network thinking. *Artific. Intell.* **170**:1194 – 212.

[33] Newth, D., Finnigan, J. (2006) Emergence and self-organization in chemistry and biology. *Aust. J. Chem.* **59**:841 – 8.

[34] Markose, S.M. (2005) Computability and evolutionary complexity: Markets as complex adaptive systems (CAS). *Econom. J.* **115**:F159-F92.

[35] Amaral, L.A.N., Ottino, J.M. (2004) Complex networks – Augmenting the framework for the study of complex systems. *Eur. Phys. J. B.* **38**:147 – 62.

[36] Gisiger, T. (2001) Scale invariance in biology: coincidence or footprint of a universal mechanism? *Biol. Rev.* **76**:161 – 209.

[37] Hess, B. (2000) Periodic patterns in biology. *Naturwiss.* **87**:199 – 211.

[38] Wales, D.J., Scheraga, H.A. (1999) Review: Chemistry – Global optimization of clusters, crystals, and biomolecules. *Science* **285**:1368 – 72.

[39] Halley, J.D., Winkler, D.A. (2006) Classification of self-organization and emergence in chemical and biological systems. *Aust. J. Chem.* **59**:849 – 53.

[40] Cramer, J.A., Booksh, K.S. (2006) Chaos theory in chemistry and chemometrics: a review. *J. Chemom.* **20**:447 – 54.

[41] Halley, J.D., Winkler, D.A. (2008) Consistent concepts of self-organization and self-assembly. *Complexity* **14**(2):10 – 17.

[42] Kenkel, N.C., Walker, D.J. (1996) Fractals in the Biological Sciences. *COENOSES* **11**:77 – 100.

[43] Mandelbrot, B.B. (1977) Fractals: Form, Chance and Dimension. San Francisco: W. H. Freeman.

[44] Mandelbrot, B.B. (1982) The fractal geometry of nature. San Francisco: W. H. Freeman.

[45] Sornette, D. (2000) Critical Phenomena in Natural Sciences. Chaos, Fractals, Self-Organization and Disorder: Concepts and Tools. Berlin: Springer.

[46] Avnir, D., Bihan, O., Malcai, O. (1998) Is the geometry of nature fractal? *Nature* **279**:39 – 40.

[47] Schroeder, M. (1991) Fractals, Chaos, Power Laws: Minutes from an Infinite Paradise. New York: W. H. Freeman and Company.

[48] Hughes, D., Paczuski, M. (2002) Large Scale Structures, Symmetry, and Universality. *Phys Rev Lett.* **88**:054302.

[49] Ward, M. (2001) Universality: the underlying theory behind life, the universe and everything. London: Pan Books.

[50]   Stanley, H.E., Amaral, L.A.N., Gopikrishnan, P., Ivanov, P.C., Keitt, T.H., Plerou, V. (2000) Scale invariance and universality: organizing principles in complex systems. *Physica A.* **281**:60 – 8.

[51]   Ryan, A.J. (2007) Emergence is coupled to scope, not level. *Complexity* **13**:67 – 77.

[52]   Halley, J.D., Winkler, D.A. (2008) A simple description of emergence and its relation to self-organization. *Complexity* **13**:10 – 15.

[53]   Coleman, P. (2007) Frontier at your fingertips. *Nature* **446**:379.

[54]   Hopkins, A.L. (2007) Network pharmacology. *Nat. Biotechnol.* **25**:1110 – 1.

[55]   Hopkins, A.L., Mason, J.S., Overington, J.P. (2006) Can we rationally design promiscuous drugs? *Curr. Opin. Struct. Biol.* **16**(1):127 – 136.

[56]   Kitano, H. (2007) A robustness-based approach to systems-oriented drug design. *Nat. Rev. Drug Disc.* **6**:202 – 10.

[57]   Sharom, J.R., Bellows, D.S., Tyers, M. (2004) From large networks to small molecules. *Curr. Opin. Chem. Biol.* **8**:81 – 90.

[58]   Winkler, D.A. (2008) Network models in drug discovery and regenerative medicine. *Biotech. Ann. Rev.* **14**:143 – 170.

[59]   Zaikin, A.N., Zhabotin, A.M. (1970) Concentration Wave Propagation in 2-Dimensional Liquid-Phase Self-Oscillating System. *Nature* **225**:535.

[60]   Maeda, S., Hara, Y., Sakai, T., Yoshida, R., Hashimoto, S. (2007) Self-walking gel. *Adv. Mat.* **19**:3480.

[61]   Yashin, V.V., Balazs, A.C. (2006) Pattern Formation and Shape Changes in Self-Oscillating Polymer Gels. *Science* **314**:798 – 801.

[62]   Shalaev, V.M., Markel, V.A., Poliakov, E.Y., Armstrong, R.L., Safonov, V.P., Sarychev, A.K. (1998) Nonlinear Optical Phenomena in Nanostructured Fractal Materials. *J. Nonlin. Opt. Phys. Mat.* **7**:131 – 52.

[63]   Hibbert, D.B., Melrose, J.R. (1988) Copper electrodeposits in paper support. *Phys. Rev. A.* **38**:1036 – 48.

[64]   de Greef, T.F.A., Meijer, E.W. (2008) Materials science: Supramolecular polymers. *Nature* **453**:171 – 3.

[65]   Kauffman, S.A. (1995) At Home in the Universe. The Search for Laws of Self-Organization and Complexity. New York: Oxford University Press.

[66]   Wintner, E.A., Conn, M.M., Rebek, J. (1994) Self-Replicating Molecules: A Second Generation. *J. Am. Chem. Soc.* **116**:8877 – 84.

[67]   Paul, N., Joyce, G.F. (2004) Minimal self-replicating systems. *Curr. Opin. Chem.-Biol.* **8**:634 – 9.

[68]   Benner, S.A., Sismour, A.M. (2005) Synthetic biology. *Nat. Rev. Genet.* **6**:533 – 43.

[69]   Cornish-Bowden, A. (2006) Putting the Systems back into Systems Biology. *Persp. Biol. Med.* **49**:475 – 89.

[70]   Rosen, R. (1991) Life itself:A comprehensive inquiry into the nature, origin, and fabrication of life. New York: Columbia Univ.Press.

[71]   Corbett, P.T., Leclaire, J., Vial, L., West, K.R., Wietor, J.-L., Sanders, J.K.M., *et al.* (2006) Dynamic Combinatorial Chemistry. *Chem. Rev.* **106**:3652 – 711.

[72]   Tan, D.S. (2005) Diversity-oriented synthesis: exploring the intersections between chemistry and biology. *Nat. Chem. Biol.* **1**:74 – 84.

[73]   Neri, F., Toivanen, J., Makinen, R.A.E. (2007) An adaptive evolutionary algorithm with intelligent mutation local searchers for designing multidrug therapies for HIV. *Applied Intelligence* **27**(3):219 – 35.

[74]   Grosdidier, A., Zoete, V., Michielin, O. (2007) EADock: Docking of small molecules into protein active sites with a multiobjective evolutionary optimization. *Proteins* **67**(4):1010 – 25.

[75]   Jain, A.N. (2006) Scoring functions for protein-ligand docking. *Curr. Prot. Pept. Sci.* **7**:407 – 20.

[76]   Jung, Y.S., Kulshreshtha, C., Kim, J.S., Shin, N., Sohn, K.S. (2007) Genetic algorithm-assisted combinatorial search for new blue phosphors in a (Ca,Sr,Ba, Mg,Eu)(x)ByPzO delta system. *Chem. Mat.* **19**:5309 – 18.

[77]   Serra JM, Corma A, Valero S, Argente E, Botti V.(2007) Soft computing techniques applied to combinatorial catalysis: A new approach for the discovery and optimization of catalytic materials. *QSAR Combin. Sci.* **26**:11 – 26.

[78]   Belda, I., Madurga, S., Tarrago, T., Llora, X., Giralt, E. (2007) Evolutionary computation and multimodal search: A good combination to tackle molecular diversity in the field of peptide design. *Mol. Divers.* **11**:7 – 21.

[79]   Hohm, T., Limbourg, P., Hoffmann, D. (2006) A multiobjective evolutionary method for the design of peptidic mimotopes. *J. Comp. Biol.* **13**:113 – 25.

[80]   Nicolaou, C.A., Brown, N., Pattichis, C.S. (2007) Molecular optimization using computational multi-objective methods. *Curr. Opin. Drug Disc. Dev.* **10**:316 – 24.

[81]   Rusu, T., Bulacovsch, V. (2006) Multiobjective tabu search method used in chemistry. *Int. J. Quant. Chem.* **106**:1406 – 12.

[82] Lameijer, E.W., Kok, J.N., Back, T., Ijzerman, A.P. (2006) The molecule evoluator. An interactive evolutionary algorithm for the design of drug-like molecules. *J. Chem. Info. Mod.* **46**:545 – 52.

[83] Liao, C.Z., Liu, B., Shi, L.M., Zhou, J.J., Lu, X.P. (2005) Construction of a virtual combinatorial library using SMILES strings to discover potential structure-diverse PPAR modulators. *Eur. J. Med. Chem.* **40**:632 – 40.

[84] Gillet, V.J., Khatib, W., Willett, P., Fleming, P.J., Green, D.V.S. (2002) Combinatorial library design using a multiobjective genetic algorithm. *J. Chem. Inf. Comp. Sci.* **42**:375 – 85.

[85] Weber, L. (2002) Multi-component reactions and evolutionary chemistry. *Drug Dis. Today* **7**:143 – 7.

[86] Yildirim, M.A., Goh, K., Cusick, M.E., Barabasi, L., Vidal, M. (2007) Drug-target network. *Nat. Biotechnol.* **25**:1119 – 26.

[87] Whitesides, G.M., Boncheva, M. (2002) Beyond molecules: self-assembly of meso-scopic and macroscopic components. *Proc. Nat. Acad. Sci. U.S.A* **99**:4769 – 74.

[88] Lehn, J.-M. (1995) Supramolecular Chemistry. New York Weinheim.

[89] Lindoy, L.F., Atkinson, I.M. (2000) Self-assembly in Supramolecular Systems. Cambridge, UK: Royal Society of Chemistry.

[90] Whitesides, G.M., Ferguson, G.S., Allara, D., Scherson, D., Speaker, L., Ulman, A. (1993) Organized molecular assemblies. *Crit. Rev. Surf. Chem.* **3**:49 – 65.

[91] Black, C.T., Ruiz, R., Breyta, G., Cheng, J.Y., Colburn, M.E., Guarini, K.W., *et al.* (2007) Polymer self assembly in semiconductor microelectronics. *IBM J. Res. Dev.* **51**:605 – 33.

[92] Sitharam, M., Agbandje-Mckenna, M. (2006) Modelling virus self-assembly pathways: Avoiding dynamics using geometric constraint decomposition. *J. Comp. Biol.* **13**:1232 – 65.

[93] Fukushima, T., Jin, W., Aida, T. (2007) Graphitic nanotubes formed by programmed self-assembly. *J. Synth. Org. Chem. Jap.* **65**:852 – 61.

[94] Arumugam, P., Xu, H., Srivastava, S., Rotello, V.M. (2007) 'Bricks and mortar' nanoparticle self-assembly using polymers. *Polymer Int.* **56**:461 – 6.

[95] Shimizu, T. (2006) Self-assembly in discrete organic nanotubes. *Sen-I Gakkaishi* **62**:P114-P8.

[96] Corti, M., Cantu, L., Brocca, P., Del Favero, E. (2007) Self-assembly in glycolipids. *Curr. Opin. Coll. Inter. Sci.* **12**:148 – 54.

[97]   Colombo, G., Soto, P., Gazit, E. (2007) Peptide self-assembly at the nanoscale: a challenging target for computational and experimental biotechnology. *Trends Biotech.* **25**:211 – 8.

[98]   Kokkoli, E., Mardilovich, A., Wedekind, A., Rexeisen, E.L., Garg, A., Craig, J.A. (2006) Self-assembly and applications of biomimetic and bioactive peptide-amphiphiles. *Soft Matter* **2**:1015 – 24.

[99]   Gray, D.G., Roman, M. (2006) Self-assembly of cellulose nanocrystals: Parabolic focal conic films. In: *Cellulose Nanocomposites: Processing, Characterization, and Properties*, (K. Oksman and M. Sain, Eds.) ACS Symposium Series 938, American Chemical Society, Washington DC, U.S.A., p. 26 – 32.

[100] Wang, X.S., Winnik, M.A., Manners, I. (2006) Synthesis, self-assembly, and applications of polyferrocenylsilane block copolymers. In: *Metal-Containing and Metallo-supramolecular Polymers and Materials*, (U.S. Schubert, G.R. Newkome, I. Manners, Eds.) ACS Symposium Series 928. American Chemical Society, Washington DC, p. 274 – 91.

[101] Johnston, M.R., Latter, M.J. (2005) Capsules, cages and three-dimensional hosts: Self-assembly of complementary monomers. *Supramol. Chem.* **17**:595 – 607.

[102] Dumoulin, M., Kumita, J.R., Dobson, C.M. (2006) Normal and aberrant biological self-assembly: Insights from studies of human lysozyme and its amyloidogenic variants. *Acc. Chem. Res.* **39**:603 – 10.

[103] Gazit, E. (2005) Mechanisms of amyloid fibril self-assembly and inhibition. *FEBS J.* **272**:5971 – 8.

[104] Whitesides, G.M., Grzybowski, B. (2002) Self-assembly at all scales. *Science* **295**:2418 – 21.

[105] Pfeifer, R., Lungarella, M., Iida, F. (2007) Self-organization, embodiment, and biologically inspired robotics. *Science* **318**:1088 – 93.

[106] Miller, A.D. (2002) Order for free: Molecular diversity and complexity promote self-organisation. *Chembiochem* **3**:45 – 6.

[107] Gerhart, J., Kirschner, M. (1997) Cells, Embryos, and Evolution. MA: Blackwell Science.

[108] Halley, J.D., Winkler, D.A. (2008) Critical-like self-organization and natural selection: two facets of a single evolutionary process? *Biosystems* **92**:148 – 158.

[109] Lehn, J.-M. (2007) From supramolecular chemistry towards constitutional dynamic chemistry and adaptive chemistry. *Chem. Soc. Rev.* **36**:151 – 60.

[110] Figueiredo, M.A.T. (2003) Adaptive sparseness for supervised learning. *IEEE Trans. Patt. Anal. Mach. Intell.* **25**:1150 – 9.

[111] Winkler, D. (2001) The broader applications of neural and genetic modelling methods. *Drug Disc. Today* **6**:1198 – 9.

[112] Winkler, D.A. (2004) Neural networks in ADME and toxicity prediction. *Drugs Future* **29**:1043 – 57.

[113] Winkler, D.A. (2004) Neural networks as robust tools in drug lead discovery and development. *Mol. Biotech.* **27**:139 – 67.

[114] Winkler, D.A., Burden, F.R. (2000) Robust QSAR models from novel descriptors and Bayesian Regularised Neural Networks. *Mol. Sim.* **24**:243.

[115] Winkler, D.A., Burden, F.R. (2002) Application of neural networks to large dataset QSAR, virtual screening and library design. In: *Combinatorial Chemistry Methods and Protocols* (Ed. Bellavance-English, L.). Humana Press.

[116] Winkler, D.A., Burden, F.R. (2004) Bayesian Neural Networks for Modelling in Drug Discovery. *Biosilico* **2**:104 – 11.

[117] Polley, M.J., Winkler, D.A., Burden, F.R. (2004) Broad-based quantitative structure-activity relationship modeling of potency and selectivity of farnesyltransferase inhibitors using a Bayesian regularized neural network. *J. Med. Chem.* **47**:6230 – 8.

[118] Wang, H., Chen, B., Yao, S.Z. (2006) Quantitative structure-activity relationship modelling of angiotensin converting enzyme inhibitors by back propagation artificial neural network. *Chin. J. Anal. Chem.* **34**:1674 – 8.

[119] Fernandez, M., Carreiras, M.C., Marco, J.L., Caballero, J. (2006) Modelling of acetylcholinesterase inhibition by tacrine analogues using Bayesian-regularized Genetic Neural Networks and ensemble averaging. *J. Enz. Inhib. Med. Chem.* **21**:647 – 61.

[120] Fernandez, M., Caballero, J. (2006) Bayesian-regularized genetic neural networks applied to the modelling of non-peptide antagonists for the human luteinizing hormone-releasing hormone receptor. *J. Mol. Graph. Mod.* **25**:410 – 22.

[121] Arakawa, M., Hasegawa, K., Funatsu, K. (2006) QSAR study of anti-HIV HEPT analogues based on multi-objective genetic programming and counter-propagation neural network. *Chemom. Intell. Lab. Sys.* **83**:91 – 8.

[122] Polley, M.J., Burden, F.R., Winkler, D.A. (2005) Predictive human intestinal absorption QSAR models using Bayesian regularized neural networks. *Aust. J. Chem.* **58**:859 – 63.

[123] Winkler, D.A., Burden, F.R. (2004) Modelling blood-brain barrier partitioning using Bayesian neural nets. *J. Mol. Graph. Mod.* **22**:499 – 505.

[124] Jung, E., Kim, J., Kim, M., Jung, D.H., Rhee, H., Shin, J.M. Choi, K., Kang, S.K., Kim, M.K., Yun, C.H., Choi, Y.J., Choi, S.H. (2007) Artificial neural network models for prediction of intestinal permeability of oligopeptides. *BMC Bioinformatics* **8**:245.

[125] Di Fenza, A., Alagona, G., Ghio, C., Leonardi, R., Giolitti, A., Madami, A. (2007) Caco-2 cell permeability modelling: a neural network coupled genetic algorithm approach. *J. Comp.-Aided Mol. Des.* **21**:207 – 21.

[126] Chen, L.J., Lian, G.P., Han, L.J. (2007) Prediction of human skin permeability using artificial neural network (ANN) modelling. *Acta Pharm. Sinica.* **28**:591 – 600.

[127] Doytchinova, I.A., Flower, D.R. (2003) Towards the in silico identification of class II restricted T-cell epitopes: a partial least squares iterative self-consistent algorithm for affinity prediction. *Bioinfo.* **19**:2263 – 70.

[128] Winkler, D.A., Burden, F.R. (2005) Predictive Bayesian Neural Network Models of MHC Class II Peptide Binding. *J. Mol. Graph. Mod.* **23**:481 – 9.

[129] Batouche, S., Rebbani, N., Gheid, A. (2006) Artificial neural network and topological indices to predict retention indices in gas chromatography. *Asian J. Chem.* **18**:2623 – 36.

[130] Liu, G.S., Yu, J.G. (2005) QSAR analysis of soil sorption coefficients for polar organic chemicals: Substituted anilines and phenols. *Water Res.* **39**:2048 – 55.

[131] Tetko, I.V., Bruneau, P. (2004) Application of ALOGPS to predict 1-octanol/water distribution coefficients, logP, and logD, of AstraZeneca in-house database. *J. Pharm. Sci.* **93**:3103 – 10.

[132] Eros, D., Kovesdi, I., Orfi, L., Takacs-Novak, K., Acsady, G., Keri, G. (2002) Reliability of logP predictions based on calculated molecular descriptors: A critical review. *Curr. Med. Chem.* **9**:1819 – 29.

[133] Arupjyoti, S., Iragavarapu, S. (1998) New electrotopological descriptor for prediction of boiling points of alkanes and aliphatic alcohols through artificial neural network and multiple linear regression analysis. *Comp. Chem.* **22**:515 – 22.

[134] Zhang, R.S., Liu, S.H., Liu, M.C., Hu, Z. (1997) Neural network molecular descriptors approach to the prediction of properties of alkenes. *Comp. Chem.* **21**:335 – 41.

[135] Yan, A.X., Gasteiger, J., Krug, M., Anzali, S. (2004) Linear and nonlinear functions on modeling of aqueous solubility of organic compounds by two structure representation methods. *J. Comp.-Aided Mol. Des.* **18**:75 – 87.

[136] Huuskonen, J., Rantanen, J., Livingstone, D. (2000) Prediction of aqueous solubility for a diverse set of organic compounds based on atom-type electrotopological state indices. *Eur. J. Med. Chem.* **35**:1081 – 8.

[137] Burden, F.R., Winkler, D.A. (2000) A quantitative structure-activity relationships model for the acute toxicity of substituted benzenes to *Tetrahymena pyriformis* using Bayesian-regularized neural networks. *Chem. Res. Toxicol.* **13**:436 – 40.

[138] Melagraki, G., Afantitis, A., Sarimveis, H., Igglessi-Markopoulou, O., Alexandridis, A. (2006) A novel RBF neural network training methodology to predict toxicity to *Vibrio fischeri*. *Mol. Divers.* **10**:213 – 21.

[139] Caballero, J., Garriga, M., Fernandez, M. (2005) Genetic neural network modelling of the selective inhibition of the intermediate-conductance Ca^{2+}-activated K^+ channel by some triarylmethanes using topological charge indexes descriptors. *J. Comp.-Aided Mol. Des.* **19**:771 – 89.

[140] Vracko, M., Mills, D., Basak, S.C. (2004) Structure-mutagenicity modelling using counter propagation neural networks. *Environ. Toxicol. Pharmacol.* **16**:25 – 36.

[141] Mazzatorta, P., Vracko, M., Jezierska, A., Benfenati, E. (2003) Modeling toxicity by using supervised Kohonen Neural Networks. *J. Chem. Info. Comp. Sci.* **43**:485 – 92.

[142] Admans, G., Takahashi, Y., Ban, S., Kato, H., Abe, M., Hanai, S. (2001) Artificial neural network for predicting the toxicity of organic molecules. *Bull. Chem. Soc. Jap.* **74**:2451 – 61.

[143] Keller, E.F. (2005) Revisiting "scale free" networks. *Bioessays* **27**:1060 – 1068.

[144] Geard, N., Wiles, J. (2005) A gene network model for developing cell lineages. *Artif. Life* **11**:249 – 67.

[145] Winkler, D.A., Burden, F.R., Halley, J.D. (2008) Using recursive networks to describe and predict gene expression and fate decisions during *C. elegans* embryogenesis. *Artif. Life* in press.

[146] Weaver, D.C. (1999) Modelling regulatory networks with weight matrices. *Pac. Symp. Biocomp.* **4**:112 – 23.

[147] Hussain, M.A., Theise, N.D. (2004) Post-natal stem cells as participants in complex systems and the emergence of tissue integrity and function. *Pediatric Diabetes* **5**:75 – 8.

[148] Laslett, A.L., Grimmond, S., Gardiner, B., Stamp, L., Lin, A., Hawes, S.M., *et al.* (2007) Transcriptional Analysis of Early Lineage Commitment In Human Embryonic Stem Cells. *BMC Dev. Biol.* **7**:12 – 30.

[149] Welch, J.J., Watts, J.A., Vakoc, C.R., Yao, Y. (2004) Global regulation of erythroid gene expression by transcription factor GATA-1. *Blood* **104**:3136–47.

[150] Hamatani, T., Carter, M.G., Sharov, A.A., Ko, M.S.H. (2004) Dynamics of Global Gene Expression Changes during Mouse Preimplantation Development. *Dev. Cell* **6**:117–31.

[151] Hart, C.E., Mjolsness, E., Wold, B.J. (2006) Transcription network: Inferences from neural networks. *PloS Comp. Biol.* **2**:1592–607.

 Beilstein-Institut

Systems Chemistry, May 26th – 30th, 2008, Bozen, Italy

The Cellular Uptake of Pharmaceutical Drugs is Mainly Carrier-mediated and is Thus an Issue not so Much of Biophysics but of Systems Biology

Douglas B. Kell[*] and Paul D. Dobson

School of Chemistry and Manchester Interdisciplinary Biocentre,
The University of Manchester,
131 Princess St, Manchester M1 7DN, UK

E-Mail: *dbk@manchester.ac.uk

Received: 8th August 2008 / Published: 16th March 2009

Abstract

It is widely believed that most drug molecules are transported across the phospholipid bilayer portion of biological membranes via passive diffusion at a rate related to their lipophilicity (expressed as log P, a calculated c log P or as log D, the octanol:water partition coefficient). However, studies of this using purely phospholipid bilayer membranes have been very misleading since transfer across these typically occurs via the solvent reservoirs or via aqueous pore defects, neither of which are prevalent in biological cells. Since the types of biophysical forces involved in the interaction of drugs with lipid membranes are no different from those involved in their interaction with proteins, arguments based on lipophilicity also apply to drug uptake by membrane transporters or carriers. A similar story attaches to the history of mechanistic explanations of the mode of action of general anaesthetics (narcotics). Carrier-mediated and active uptake of drugs is far more common than is usually assumed. This has considerable implications for the design of libraries for drug discovery and development, as well as for chemical genetics/genomics and systems chemistry.

INTRODUCTION

As is well known (e. g. [1 – 4]), attrition rates of drugs in pharmaceutical companies remain extremely high, and nowadays this is mainly due either to lack of efficacy or for reasons of toxicity. Arguably these issues are mainly due to the fact that drug candidates are typically isolated on the basis of their potency in a screen against a molecular target, and only subsequently are they tested in organisms *in vivo*. Since most modern targets have enjoyed some degree of validation using e. g. genetic knockouts, it is likely that the problem of ostensible potency *in vitro* but lack of efficacy *in vivo* is not so much with the target but with the ability of the drug to find the target. In a similar vein, if drugs are accumulated to high levels in particular tissues via the action of active solute transporters [5 – 7], it is the cellular and tissue distributions of the relevant carriers, rather than any general biophysical properties of the drugs of interest, that largely determine differential tissue distributions. An overview of this article is given in Figure 1. We begin by rehearsing some of the relevant arguments.

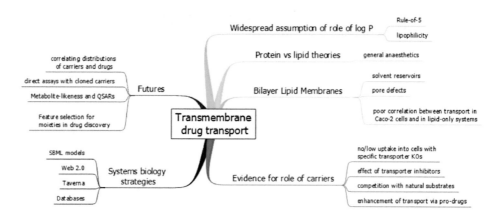

Figure 1. An overview of this article in the form of a 'mind map' [8].

HOW DRUGS CROSS CELL MEMBRANES

The prevailing view of cell membranes, popularised in Singer and Nicolson's celebrated paper of 1972 [9], is that of polytopic proteins floating (and diffusing) in a 'sea' of phospholipid bilayer, as illustrated in the cartoon of Figure 2.

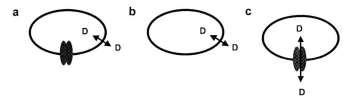

Figure 2. A simple cartoon of two means by which a molecule such as a drug (D) may cross a cellular membrane, either by diffusing through the phospholipid bilayer portion (**a, b**) or being taken up via a carrier (**c**) (or both).

While the main elements of this are broadly accepted, two features are of note. First, the protein:lipid ratio in membranes (by mass) is typically 1:1 and may be 3:1 [10], and secondly that most lipids are partially or significantly influenced by the presence of the protein component (and vice versa [11, 12]). However, the cartoon serves to cover the nexus of this article, viz. the question of whether drugs mainly cross cellular membranes via passage through the phospholipid bilayer portion or using carrier-mediated transport. Because it may be active, i.e. coupled to sources of free energy, the latter in particular, *modulo* the existence of any membrane potential differences between compartments, is capable of effecting considerable concentrative uptake. The question then arises as to whether there are molecular or biophysical properties of drugs that can serve to explain their rate of transfer across biological membranes.

'LIPOPHILICITY' AS A CANDIDATE DESCRIPTOR FOR RATES OF DRUG TRANSPORT ACROSS BIOMEMBRANES

From the time of Overton [13] it has been recognised that the transmembrane permeability of non-electrolytes correlates well with their olive oil (nowadays octanol): water partition coefficients, typically referred to as log D or log P (A more recent example with data can be found in [14]). Thus there has been a tendency to assume that this gives a mechanistic explanation by which such solutes must 'dissolve' or partition into the bilayer portion of such biological membranes in order to cross them. Actually it means no such thing, as the biophysical forces and mechanistic acts (e.g. making and breaking of H-bonds) required for 'partitioning' into appropriately hydrophobic protein pockets are the same, and so such correlations may also mean that solute transfer is protein-mediated (and see below).

LIPINSKI'S "RULE OF FIVE" FOR DESCRIBING DRUG BIOAVAILABILITY

As indicated above, drugs will only work when they can reach and thereby interact with their 'targets', and a first step in understanding this relates to their so-called 'bioavailability' [15 – 17], a term that covers (among others things) solubility, absorption and permeability. Indeed, it was the need to understand bioavailability that led Lipinski to devise his famous 'rule of five' (Ro5) [18]. The Ro5 predicts that poor absorption or permeation is more likely when there are more than 5 H-bond donors, 10 H-bond acceptors, the molecular weight (MW) is greater than 500 and the calculated Log P (CLogP) is greater than 5. While empirical, the Ro5 has been massively important in influencing thinking about the kinds of molecules companies might which to consider in designing drug screening libraries and the subsequent drugs [4, 19–21]. Clearly it recognises the need to balance the forces that enable a molecule to be at once both sufficiently hydrophilic to dissolve adequately in aqueous media with a requirement to be sufficiently lipophilic to penetrate to or via more hydrophobic environments. It was also explicitly recognised [18] that the Ro5 did not apply to carrier-mediated uptake, and that many/most natural products 'disobey' the Ro5 (In the more recently developed fragment-based screening – see e.g. [22, 23] – there is an even more stringent 'rule of three' [24]). Log P in its various incarnations is thus seen as a very important property of a candidate drug molecule, although as a macroscopic property it is not entirely obvious how this would be terribly predictive of drug distributions.

BILAYER OR BLACK LIPID MEMBRANES

Notwithstanding this, the long history of the relation between permeability and log P, coupled to the implication that it simply involves dissolving in a hydrophobic environment while crossing from one aqueous phase to another *in vivo*, has meant that many have sought to simplify the understanding of transmembrane molecule transport by studying it in bilayer or 'black' lipid membranes (BLMs) [25–28] lacking protein (Fig. 3).

Black (Bilayer) Lipid Membranes (BLMs)

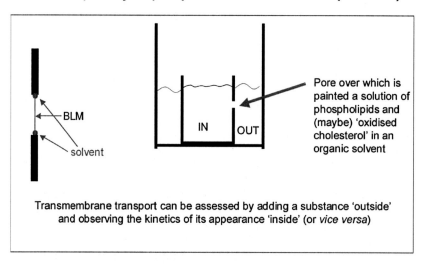

Figure 3. The principle of formation of conventional BLMs

The problems with this kind of system are (i) that most BLMs are formed using organic solvents, and the residual solvent reservoirs (forming an annulus as the edge of the BLM, see Fig. 3) provide a vehicle for transport that does not involve the phospholipid bilayer, and (ii) that many BLMs exhibit aqueous pore defects that biomembranes do not (see also [29, 30]), some potentially induced by solutes themselves, and that these permit transport that does not therefore involve dissolution in any phospholipid (e.g. [31 – 38]). Indeed, the enormous Born charging energy for transferring electrostatic charge across any low dielectric medium is prohibitive to the trans-lipid transport of ionic charges [39, 40]. Consequently, it is rather doubtful whether such model systems possess the properties necessary for them to act as a useful guide for the mechanisms of transport via natural membranes. The rate of transport of drugs across more recently devised lipid-only membrane systems is also only weakly correlated with the transport of the same molecules across biological membranes (see e.g. [7, 41 – 44]).

THE MECHANISM OF ACTION OF GENERAL ANAESTHETICS (NARCOTICS)

Correlation is of course a poor guide to mechanism or causality, and another example where there are excellent correlations between bioactivity and lipophilicity, but where these have proved mechanistically highly misleading, is represented by the mode of action of narcotic agents ('general anaesthetics'). Starting with the studies of Meyer [45] and of Overton [46] (see also [47]), a close relationship between lipophilicity (lop D) and narcotic potency was established. The almost complete lack of a structure-activity relationship over 5 orders of magnitude (but cf. [48, 49]) led many to assume that a simple biophysically based partitioning of anaesthetic molecules into cell membranes (followed, presumably, by some kind of inhibitory pressure-induced effect on membrane ion channels) could account for narcosis [50] (and also tended to imply a unitary mechanism). However, a number of molecules deviate considerably from this picture, and some isomers with similar biophysical properties have very different anaesthetic potencies [51 – 53]. Now, the biophysical forces underpinning the interaction of such molecules with lipids are no different from those describing interactions with proteins [54], and indeed equivalent interactions of these molecules with fully soluble (non-membranous) proteins (e.g. [55 – 58], including direct structural evidence for binding [57, 59], and the correlation between specific receptor binding (e.g. [60]) and potency in specific mutant mice [61] (and see [62]), mean that this view is no longer considered tenable (e.g. [54, 63 – 70]), and it is now recognised that general anaesthetics of different functional classes have a variety of proteinaceous targets [54, 71], in particular GABA$_A$ receptor subtypes [72, 73]. Indeed, even such a small molecule as ethanol is now recognised as having relatively specific receptors [74]! Lipophilicity, and a gross analysis of chemical structure *per se*, then, are poor guides to mechanism.

EVIDENCE THAT DRUGS DO HITCHHIKE ON 'NATURAL' CARRIERS

Space does not permit an exhaustive review, and printed papers as such are a poor means to summarise knowledge of this type [75]. However, following early indications that even lipophilic cations require carriers for transmembrane transport [76], a huge number of 'exceptions' (or at least instances) have been found. Some are listed in the supplementary information to our recent review [7] while others can be found in other summaries [5, 6]. To this end, we shall shortly be making available a database of human drug transporters (see also [77]), based in part on our data model for metabolite databases [78, 79].

SYSTEMS BIOLOGY, DATABASES AND WEB 2.0 FOR UNDERSTANDING DRUG UPTAKE

There is now a convergence [80 – 82] between (i) our understanding of those human meta-bolites that can be determined from genomic reconstructions and the literature [83 – 85] and (ii) the metabolic network models that alone will allow to effect true systems biology modelling [86 – 89]. Our main strategy for assisting this involves the use of workflows of loosely coupled elements [80 – 82, 90], with the models encoded in SBML [91] (www.sbml.org) according to principled markup standards [92, 93]. Bottom-up approaches (Fig. 4) have the merit of starting with molecular mechanism, but do rely on knowledge of the relevant participants.

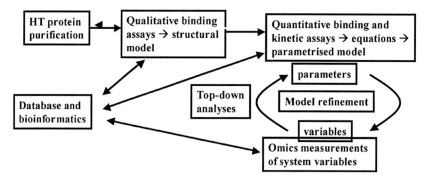

Figure 4. A 'bottom-up' systems biology approach (including top-down strategies, and thereby 'middle-out' [94]) with which to develop metabolic network models that include drug transporters

WHERE NEXT?

The human genome encodes some 900+ drug transporters [95, 96], and while the main ones involved in cellular drug uptake are a comparatively small subset of these [7], it is clear that we need to understand the specificity and distribution of these, just as is the case with the cytochromes P450 that are so important in drug metabolism. Studies of specificity and

QSAR measurements will require comparison of drug transport into cells containing or lacking cloned carriers. The tissue and even subcellular distributions of carriers (e.g. in mitochondria [97]) are emerging from studies such as the Human Proteome Atlas (http://proteinatlas.org/) (e.g. [98, 99]. An example, showing the extreme differences in carrier expression between tissues that can be observed, is given in figure 5. Such quantitative proteomic data will be extremely valuable in assisting us in the development of systems biology models, since although it is possible to make substantial progress by 'guessing' kinetic parameters from the topology and stoicheiometry of metabolic networks alone [100], or better inferring them from measured fluxes and concentrations (e.g. [101 – 107]), experimental measurements of K_m, k_{cat} and protein concentrations is altogether more satisfactory for building and constraining kinetic models.

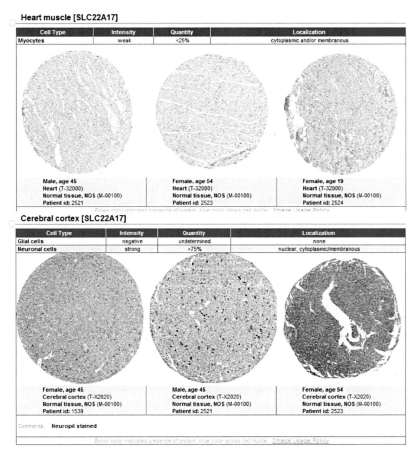

Figure 5. An example from the Human Protein Atlas, taken with permission on its website, of representative tissue distributions of the protein SLC22A17, a so-called brain-specific organic cation transporter. Links are via http://proteinatlas.org/tissue_profile.php? antibody_id = 2728.

We need to understand much better than we do now the biophysical, chemical and molecular descriptors that are important in determining drug dispositions, and this requires the production of suitable models [96]. Evolutionary computing methods (e.g. [108 – 110]) are extremely powerful but surprisingly underutilised for these purposes. Log D measurements are still of value as a 'baseline', but tend to be poor predictors even of gross biological effects when the chemical involved are not in homologous series [111].

FRAGMENT-BASED DRUG DISCOVERY AND DRUG TRANSPORTERS

One interesting approach to drug discovery, rather akin to an evolutionary computing type of approach, involves the evolution of drug structures from smaller fragments (e.g. [23, 24, 112 – 130]), and the obvious question arises as to which kinds of fragments might best be included in the libraries used. Clearly it will be of interest to compare the similarity of such fragments to natural metabolites [131], since those that are most similar to 'natural' metabolites that are known to enter cells are most likely to serve as transporter substrates (the principle of molecular similarity [132 – 135]).

CONCLUDING REMARKS

"When one admits that nothing is certain one must, I think, also admit that some things are much more nearly certain than others." [136]

One cannot fail to remark on the huge volume and continuing growth of the scientific literature. Two and a half million peer-reviewed papers are published per year [137], with over 1 million per year in Medline alone (http://www.nlm.nih.gov/bsd/medline_cit_counts_yr_pub.html). The former equates to nearly 5 refereed scientific papers being published per minute – and in a somewhat similar vein presently 10 hours of (albeit largely non-science-related) video material are added at www.youtube.com in the same time! A consequence of this is a kind of 'balkanisation' [138] of the literature in which scientists focus solely on more detailed analyses of ever smaller parts of biology. This is clearly going to change [89], and will have to do so, as a result of computerization, the internet and the emergence of systems biology, since only a global overview can lead to general truths (inductive reasoning [139]). Only by looking at many hundreds of papers did we recognize that carrier-mediated uptake is the rule and not the exception [7]. Automation is therefore required.

Given suitably digitised literature and attendant metadata [75], we need to exploit methods such as text mining [140 – 143], conceptual associations [144 – 146] and literature-based discovery (e.g. [146 – 152] to create new knowledge.

Consequently, we hope we can look forward to the development of many computational tools that will assist chemical biologists in putting together systems biology models that describe accurately the internal biochemical mechanisms of the 'digital human' [88].

ACKNOWLEDGMENTS

Our interest in pursuing these issues has been helped considerably by grant BB/D007747/1 from the BBSRC (June 2006 – May 2008), together with attendant funding from GSK. We thank Scott Summerfield and Phil Jeffrey of GSK for their support and interest, and Karin Lanthaler and Steve Oliver for useful discussions. DBK also thanks the EPSRC and RSC for financial support, and the Royal Society/Wolfson Foundation for a Research Merit Award. This is a contribution from the BBSRC- and EPSRC-funded Manchester Centre for Integrative Systems Biology (www.mcisb.org/).

REFERENCES

[1]   Ajay (2002). Predicting drug-likeness: why and how? *Curr. Top. Med. Chem.* **2**:1273 – 86.

[2]   Hann, M.M., Oprea, T.I. (2004). Pursuing the leadlikeness concept in pharmaceutical research. *Curr. Op. Chem. Biol.* **8**:255 – 263.

[3]   Kola, I., Landis, J. (2004). Can the pharmaceutical industry reduce attrition rates? *Nat. Rev. Drug Discov.* **3**:711 – 5.

[4]   Leeson, P.D. & Springthorpe, B. (2007). The influence of drug-like concepts on decision-making in medicinal chemistry. *Nat. Rev. Drug Discov.* **6**:881 – 90.

[5]   Sai, Y., Tsuji, A. (2004). Transporter-mediated drug delivery: recent progress and experimental approaches. *Drug. Discov. Today.* **9**:712 – 20.

[6]   Sai, Y. (2005). Biochemical and molecular pharmacological aspects of transporters as determinants of drug disposition. *Drug. Metab. Pharmacokinet.* **20**:91 – 9.

[7]   Dobson, P.D., Kell, D.B. (2008). Carrier-mediated cellular uptake of pharmaceutical drugs: an exception or the rule? *Nat. Rev. Drug Discov.* **7**:205 – 220.

[8]   Buzan, T. (2002). *How to mind map*. Thorsons, London.

[9]   Singer, S.J., Nicolson, G.L. (1972). The fluid mosaic model of the structure of cell membranes. *Science* **175**:720 – 31.

[10]   Westerhoff, H.V., Kell, D.B., Kamp, F., van Dam, K. (1988). The membranes involved in proton-mediated free-energy transduction: thermodynamic implications of their physical structure. In *Microcompartmentation* (ed. D. P. Jones), pp. 115 – 154. CRC Press, Boca Raton, Fl.

[11]   Lee, A.G. (2004). How lipids affect the activities of integral membrane proteins. *Biochim. Biophys. Acta* **1666**:62 – 87.

[12] Reynwar, B.J., Illya, G., Harmandaris, V.A., Muller, M.M., Kremer, K., Deserno, M. (2007). Aggregation and vesiculation of membrane proteins by curvature-mediated interactions. *Nature* **447**:461–464.

[13] Overton, E. (1899). Über die allgemeinen osmotischen Eigenschaften der Zelle, ihre vermutliche Ursachen und ihre Bedeutung für die Physiologie *Vierteljahrsschr. Naturforsch. Ges. Zürich* **44**:88–114.

[14] Lieb, W.R., Stein, W.D. (1969). Biological membranes behave as non-porous polymeric sheets with respect to the diffusion of non-electrolytes. *Nature* **224**:240–3.

[15] Amidon, G.L., Lennernas, H., Shah, V.P., Crison, J.R. (1995). A theoretical basis for a biopharmaceutic drug classification: the correlation of in vitro drug product dissolution and in vivo bioavailability. *Pharm. Res.* **12**:413–20.

[16] Lennernäs, H., Abrahamsson, B. (2005). The use of biopharmaceutic classification of drugs in drug discovery and development: current status and future extension. *J. Pharm. Pharmacol.* **57**:273–85.

[17] Wu, C.Y., Benet, L.Z. (2005). Predicting drug disposition via application of BCS: transport/absorption/elimination interplay and development of a biopharmaceutics drug disposition classification system. *Pharm. Res.* **22**:11–23.

[18] Lipinski, C.A., Lombardo, F., Dominy, B.W., Feeney, P.J. (1997). Experimental and computational approaches to estimate solubility and permeability in drug discovery and development settings. *Adv. Drug Deliv. Rev.* **23**:3–25.

[19] Lipinski, C.A. (2000). Drug-like properties and the causes of poor solubility and poor permeability. *J. Pharmacol. Toxicol. Methods* **44**:235–49.

[20] van De Waterbeemd, H., Smith, D.A., Beaumont, K., Walker, D.K. (2001). Property-based design: optimization of drug absorption and pharmacokinetics. *J. Med. Chem.* **44**:1313–33.

[21] Owens, J., Lipinski, C. (2003). Chris Lipinski discusses life and chemistry after the Rule of Five. *Drug Discov. Today* **8**:12–16.

[22] Carr, R., Jhoti, H. (2002). Structure-based screening of low-affinity compounds. *Drug Discov Today* **7**:522–7.

[23] Hartshorn, M.J., Murray, C.W., Cleasby, A., Frederickson, M., Tickle, I.J., Jhoti, H. (2005). Fragment-based lead discovery using X-ray crystallography. *J. Med. Chem.* **48**:403–13.

[24] Congreve, M., Carr, R., Murray, C., Jhoti, H. (2003). A rule of three for fragment-based lead discovery? *Drug Discov. Today* **8**:876–877.

[25] Mueller, P., Rudin, D.O., Tien, H.T., Wescott, W.C. (1962). Reconstitution of cell membrane structure *in vitro* and its transformation into an excitable system. *Nature* **194**:979 – 980.

[26] Jain, M.K. (1972). *The bimolecular lipid membrane.* Van Nostrand Reinhold, New York.

[27] Tien, H.T. (1974). *Bilayer lipid membranes (BLM): theory and practice.* Marcel Dekker, New York.

[28] Tien, H.T., Ottova-Leitmannova, A. (2003). *Planar lipid bilayers (BLMs) and their applications.* Elsevier, New York.

[29] Deamer, D.W. (2008). Origins of life: How leaky were primitive cells? *Nature* **454**:37 – 8.

[30] Mansy, S.S., Schrum, J.P., Krishnamurthy, M., Tobe, S., Treco, D.A., Szostak, J.W. (2008). Template-directed synthesis of a genetic polymer in a model protocell. *Nature* **454**:122 – 5.

[31] Jansen, M., Blume, A. (1995). A comparative study of diffusive and osmotic water permeation across bilayers composed of phospholipids with different head groups and fatty acyl chains. *Biophys. J.* **68**:997 – 1008.

[32] Bordi, F., Cametti, C., Naglieri, A. (1999). Ion transport in lipid bilayer membranes through aqueous pores. *Coll. Surf. A* **159**:231 – 237.

[33] Leontiadou, H., Mark, A.E., Marrink, S.J. (2004). Molecular dynamics simulations of hydrophilic pores in lipid bilayers. *Biophys. J.* **86**:2156 – 64.

[34] Tieleman, D.P., Marrink, S.J. (2006). Lipids out of equilibrium: energetics of desorption and pore mediated flip-flop. *J. Am. Chem. Soc.* **128**:12462 – 7.

[35] Xiang, T.X., Anderson, B.D. (2006). Liposomal drug transport: a molecular perspective from molecular dynamics simulations in lipid bilayers. *Adv. Drug. Deliv. Rev.* **58**:1357 – 78.

[36] Gurtovenko, A.A., Vattulainen, I. (2007). Molecular mechanism for lipid flip-flops. *J. Phys. Chem. B* **111**:13554 – 9.

[37] Gurtovenko, A.A., Anwar, J. (2007). Ion transport through chemically induced pores in protein-free phospholipid membranes. *J. Phys. Chem .B* **111**:13379 – 82.

[38] Leontiadou, H., Mark, A.E., Marrink, S.J. (2007). Ion transport across transmembrane pores. *Biophys. J.* **92**:4209 – 15.

[39] Parsegian, A. (1969). Energy of an ion crossing a low dielectric membrane: solutions to four relevant electrostatic problems. *Nature* **221**:844 – 6.

[40] Weaver, J.C., Chizmadzhev, Y.A. (1996). Theory of electroporation: a review. *Bioelectrochem. Bioenerg.* **41**:135 – 160.

[41] Balimane, P.V., Han, Y.H., Chong, S.H. (2006). Current industrial practices of assessing permeability and P-glycoprotein interaction. *AAPS Journal* **8**:E1-E13.

[42] Corti, G., Maestrelli, F., Cirri, M., Zerrouk, N., Mura, P. (2006). Development and evaluation of an in vitro method for prediction of human drug absorption – II. Demonstration of the method suitability. *Eur. J. Pharm. Sci.* **27**:354 – 362.

[43] Galinis-Luciani, D., Nguyen, L., Yazdanian, M. (2007). Is PAMPA a useful tool for discovery? *J. Pharm. Sci.* **96**:2886 – 92.

[44] Avdeef, A., Bendels, S., Di, L., Faller, B., Kansy, M., Sugano, K., Yamauchi, Y. (2007). PAMPA – critical factors for better predictions of absorption. *J. Pharm. Sci.* **96**:2893 – 909.

[45] Meyer, H.H. (1899). Welche Eigenschaft der Anästhetica bedingt ihre narkotische Wirkung? *Arch. Exp. Pathol. Pharmakol.* **42**:109 – 118.

[46] Overton, C.E. (1901). *Studien über die Narkose zugleich ein Beitrag zur allgemeinen Pharmakologie.* Gustav Fischer, Jena.

[47] De Weer, P. (2000). A century of thinking about cell membranes. *Annu. Rev. Physiol.* **62**:919 – 26.

[48] Sewell, J.C., Sear, J.W. (2006). Determinants of volatile general anesthetic potency: a preliminary three-dimensional pharmacophore for halogenated anesthetics. *Anesth. Analg.* **102**:764 – 71.

[49] Eckenhoff, R., Zheng, W., Kelz, M. (2008). From anesthetic mechanisms research to drug discovery. *Clin. Pharmacol. Ther.* **84**:144 – 8.

[50] Seeman, P. (1972). The membrane actions of anesthetics and tranquilizers. *Pharmacol. Rev.* **24**:583 – 655.

[51] Dickinson, R., Franks, N.P., Lieb, W.R. (1994). Can the stereoselective effects of the anesthetic isoflurane be accounted for by lipid solubility? *Biophys. J.* **66**:2019 – 23.

[52] Dickinson, R., White, I., Lieb, W.R., Franks, N.P. (2000). Stereoselective loss of righting reflex in rats by isoflurane. *Anesthesiology* **93**:837 – 43.

[53] Bertaccini, E.J., Trudell, J.R., Franks, N.P. (2007). The common chemical motifs within anesthetic binding sites. *Anesth. Analg.* **104**:318 – 24.

[54] Franks, N.P. (2008). General anaesthesia: from molecular targets to neuronal pathways of sleep and arousal. *Nat. Rev. Neurosci.* **9**:370 – 86.

[55]   Ueda, I. (1965). Effects of diethyl ether and halothane on firefly luciferin bioluminescence. *Anesthesiology* **26**:603 – 6.

[56]   Franks, N.P., Lieb, W.R. (1984). Do general anaesthetics act by competitive binding to specific receptors? *Nature* **310**:599 – 601.

[57]   Franks, N.P., Jenkins, A., Conti, E., Lieb, W.R., Brick, P. (1998). Structural basis for the inhibition of firefly luciferase by a general anesthetic. *Biophys. J.* **75**:2205 – 11.

[58]   Miller, K.W. (2002). The nature of sites of general anaesthetic action. *Br. J. Anaesth.* **89**:17 – 31.

[59]   Szarecka, A., Xu, Y., Tang, P. (2007). Dynamics of firefly luciferase inhibition by general anesthetics: Gaussian and anisotropic network analyses. *Biophys. J.* **93**:1895 – 905.

[60]   Mihic, S.J., Ye, Q., Wick, M.J., Koltchine, V.V., Krasowski, M.D., Finn, S.E., Mascia, M.P., Valenzuela, C.F., Hanson, K.K., Greenblatt, E.P., Harris, R.A., Harrison, N.L. (1997). Sites of alcohol and volatile anaesthetic action on GABA$_A$ and glycine receptors. *Nature* **389**:385 – 9.

[61]   Jurd, R., Arras, M., Lambert, S., Drexler, B., Siegwart, R., Crestani, F., Zaugg, M., Vogt, K.E., Ledermann, B., Antkowiak, B., Rudolph, U. (2003). General anesthetic actions in vivo strongly attenuated by a point mutation in the GABA$_A$ receptor β3 subunit. *FASEB J.* **17**:250 – 2.

[62]   Heurteaux, C., Guy, N., Laigle, C., Blondeau, N., Duprat, F., Mazzuca, M., Lang-Lazdunski, L., Widmann, C., Zanzouri, M., Romey, G., Lazdunski, M. (2004). TREK-1, a K+ channel involved in neuroprotection and general anesthesia. *EMBO J.* **23**:2684 – 95.

[63]   Thompson, S.A., Wafford, K. (2001). Mechanism of action of general anaesthetics: new information from molecular pharmacology. *Curr. Opin. Pharmacol.* **1**:78 – 83.

[64]   Urban, B.W. (2002). Current assessment of targets and theories of anaesthesia. *Br. J. Anaesth.* **89**:167 – 83.

[65]   Campagna, J.A., Miller, K.W., Forman, S.A. (2003). Mechanisms of actions of inhaled anesthetics. *New England Journal of Medicine* **348**:2110 – 2124.

[66]   Franks, N.P., Honoré, E. (2004). The TREK K$_{2P}$ channels and their role in general anaesthesia and neuroprotection. *Trends Pharmacol. Sci.* **25**:601 – 8.

[67]   Rudolph, U., Antkowiak, B. (2004). Molecular and neuronal substrates for general anaesthetics. *Nat. Rev. Neurosci.* **5**:709 – 20.

[68]   Hemmings, H.C., Jr., Akabas, M.H., Goldstein, P.A., Trudell, J.R., Orser, B.A., Harrison, N.L. (2005). Emerging molecular mechanisms of general anesthetic action. *Trends Pharmacol. Sci.* **26**:503 – 10.

[69]   Grasshoff, C., Drexler, B., Rudolph, U., Antkowiak, B. (2006). Anaesthetic drugs: linking molecular actions to clinical effects. *Curr. Pharm. Des.* **12**:3665 – 79.

[70]   Franks, N.P. (2006). Molecular targets underlying general anaesthesia. *Br. J. Pharmacol.* **147 Suppl 1,** S 72 – 81.

[71]   Urban, B.W., Bleckwenn, M., Barann, M. (2006). Interactions of anesthetics with their targets: non-specific, specific or both? *Pharmacol. Ther.* **111**:729 – 70.

[72]   Solt, K., Forman, S.A. (2007). Correlating the clinical actions and molecular mechanisms of general anesthetics. *Curr. Opin. Anaesthesiol.* **20**:300 – 6.

[73]   Bonin, R.P., Orser, B.A. (2008). $GABA_A$ receptor subtypes underlying general anesthesia. *Pharmacol. Biochem. Behavior* **90**:105 – 112.

[74]   Wallner, M., Hanchar, H.J., Olsen, R.W. (2006). Low-dose alcohol actions on alpha 4 beta 3 delta $GABA_A$ receptors are reversed by the behavioral alcohol antagonist Ro15 – 4513. *Proc. Natl. Acad. Sci. U.S.A.* **103**:8540 – 8545.

[75]   Hull, D., Pettifer, S.R., Kell, D.B. (2008). Defrosting the digital library: bibliographic tools for the next generation web. *PLoS Comp. Biol.* **4**(10):e1000204.

[76]   Barts, P.W.J.A., Hoeberichts, J.A., Klaassen, A., Borst-Pauwels, G.W.F.H. (1980). Uptake of the lipophilic cation dibenzyldimethylammonium into *Saccharomyces cerevisiae*. Interaction with the thiamine transport system. *Biochim. Biophys. Acta* **597**:125 – 36.

[77]   Yan, Q., Sadée, W. (2000). Human membrane transporter database: a Web-accessible relational database for drug transport studies and pharmacogenomics. *AAPS PharmSci.* **2**:E20.

[78]   Jenkins, H., Hardy, N., Beckmann, M., Draper, J., Smith, A.R., Taylor, J., Fiehn, O., Goodacre, R., Bino, R., Hall, R., Kopka, J., Lane, G.A., Lange, B.M., Liu, J.R., Mendes, P., Nikolau, B.J., Oliver, S.G., Paton, N.W., Roessner-Tunali, U., Saito, K., Smedsgaard, J., Sumner, L.W., Wang, T., Walsh, S., Wurtele, E.S., Kell, D.B. (2004). A proposed framework for the description of plant metabolomics experiments and their results. *Nat. Biotechnol.* **22**:1601 – 1606.

[79]   Spasic, I., Dunn, W.B., Velarde, G., Tseng, A., Jenkins, H., Hardy, N.W., Oliver, S.G., Kell, D.B. (2006). MeMo: a hybrid SQL/XML approach to metabolomic data management for functional genomics. *BMC Bioinformatics* **7**:281.

[80] Kell, D.B. (2006). Systems biology, metabolic modelling and metabolomics in drug discovery and development. *Drug Disc. Today* **11**:1085–1092.

[81] Kell, D.B. (2006). Metabolomics, modelling and machine learning in systems biology: towards an understanding of the languages of cells. The 2005 Theodor Bücher lecture. *FEBS J.* **273**:873–894.

[82] Kell, D.B. (2007). Metabolomic biomarkers: search, discovery and validation. *Exp. Rev. Mol. Diagnost.* **7**:329–333.

[83] Duarte, N.C., Becker, S.A., Jamshidi, N., Thiele, I., Mo, M.L., Vo, T.D., Srvivas, R., Palsson, B.Ø. (2007). Global reconstruction of the human metabolic network based on genomic and bibliomic data. *Proc. Natl. Acad. Sci. U.S.A.* **104**:1777–1782.

[84] Ma, H., Sorokin, A., Mazein, A., Selkov, A., Selkov, E., Demin, O., Goryanin, I. (2007). The Edinburgh human metabolic network reconstruction and its functional analysis. *Mol. Syst. Biol.* **3**:135.

[85] Ma, H., Goryanin, I. (2008). Human metabolic network reconstruction and its impact on drug discovery and development. *Drug Discov. Today* **13**:2–8.

[86] Alon, U. (2006). *An introduction to systems biology: design principles of biological circuits.* Chapman and Hall/CRC, London.

[87] Palsson, B.Ø. (2006). *Systems biology: properties of reconstructed networks.* Cambridge University Press, Cambridge.

[88] Kell, D.B. (2007). The virtual human: towards a global systems biology of multiscale, distributed biochemical network models. *IUBMB Life* **59**:689–95.

[89] Kell, D.B., Mendes, P. (2008). The markup is the model: reasoning about systems biology models in the Semantic Web era. *J. Theoret. Biol.* **252**:538–543.

[90] Li, P., Oinn, T., Stoiland, S., Kell, D.B. (2008). Automated manipulation of systems biology models using libSBML within Taverna workflows. *Bioinformatics* **24**:287–289.

[91] Hucka, M., Finney, A., Sauro, H.M., Bolouri, H., Doyle, J.C., Kitano, H., Arkin, A.P., Bornstein, B.J., Bray, D., Cornish-Bowden, A., Cuellar, A.A., Dronov, S., Gilles, E.D., Ginkel, M., Gor, V., Goryanin, II, Hedley, W.J., Hodgman, T.C., Hofmeyr, J.H., Hunter, P.J., Juty, N.S., Kasberger, J.L., Kremling, A., Kummer, U., Le Novere, N., Loew, L. M., Lucio, D., Mendes, P., Minch, E., Mjolsness, E.D., Nakayama, Y., Nelson, M.R., Nielsen, P.F., Sakurada, T., Schaff, J.C., Shapiro, B.E., Shimizu, T.S., Spence, H.D., Stelling, J., Takahashi, K., Tomita, M., Wagner, J., Wang, J. (2003). The systems biology markup language (SBML): a medium for representation and exchange of biochemical network models. *Bioinformatics* **19**:524–31.

[92]   Le Novère, N., Finney, A., Hucka, M., Bhalla, U.S., Campagne, F., Collado-Vides, J., Crampin, E.J., Halstead, M., Klipp, E., Mendes, P., Nielsen, P., Sauro, H., Shapiro, B., Snoep, J.L., Spence, H.D., Wanner, B.L. (2005). Minimum information requested in the annotation of biochemical models (MIRIAM). *Nat. Biotechnol.* **23**:1509 – 15.

[93]   Herrgård, M.J., Swainston, N., Dobson, P., Dunn, W.B., Arga, K.Y., Arvas, M., Blüthgen, N., Borger, S., Costenoble, R., Heinemann, M., Hucka, M., Le Novère, N., Li, P., Liebermeister, W., Mo, M.L., Oliveira, A.P., Petranovic, D., Pettifer, S., Simeonidis, E., Smallbone, K., Spasiæ, I., Weichart, D., Brent, R., Broomhead, D.S., Westerhoff, H.V., Kýrdar, B., Penttilä, M., Klipp, E., Palsson, B.Ø., Sauer, U., Oliver, S.G., Mendes, P., Nielsen, J., Kell, D.B. (2008). A consensus yeast metabolic network obtained from a community approach to systems biology. *Nat. Biotechnol.* **26**:1155 – 1160.

[94]   Noble, D. (2006). *The music of life: biology beyond genes.* Oxford University Press, Oxford.

[95]   Anderle, P., Huang, Y., Sadée, W. (2004). Intestinal membrane transport of drugs and nutrients: genomics of membrane transporters using expression microarrays. *Eur. J. Pharm. Sci.* **21**:17 – 24.

[96]   Ekins, S., Ecker, G.F., Chiba, P., Swaan, P.W. (2007). Future directions for drug transporter modelling. *Xenobiotica* **37**:1152 – 70.

[97]   Palmieri, F. (2008). Diseases caused by defects of mitochondrial carriers: A review. *Biochim. Biophys. Acta* **1777**:564 – 78.

[98]   Persson, A., Hober, S., Uhlen, M. (2006). A human protein atlas based on antibody proteomics. *Curr. Opin. Mol. Ther.* **8**:185 – 90.

[99]   Barbe, L., Lundberg, E., Oksvold, P., Stenius, A., Lewin, E., Björling, E., Asplund, A., Ponten, F., Brismar, H., Uhlén, M., Andersson-Svahn, H. (2008). Toward a confocal subcellular atlas of the human proteome. *Mol. Cell Proteomics* **7**:499 – 508.

[100]   Smallbone, K., Simeonidis, E., Broomhead, D.S., Kell, D.B. (2007). Something from nothing: bridging the gap between constraint-based and kinetic modelling. *FEBS J.* **274**:5576 – 5585.

[101]   Mendes, P., Kell, D.B. (1998). Non-linear optimization of biochemical pathways: applications to metabolic engineering and parameter estimation. *Bioinformatics* **14**:869 – 883.

[102]   Moles, C.G., Mendes, P., Banga, J.R. (2003). Parameter estimation in biochemical pathways: a comparison of global optimization methods. *Genome Res.* **13**:2467 – 74.

[103] Hoops, S., Sahle, S., Gauges, R., Lee, C., Pahle, J., Simus, N., Singhal, M., Xu, L., Mendes, P., Kummer, U. (2006). COPASI: a COmplex PAthway SImulator. *Bioinformatics* **22**:3067 – 74.

[104] Rodriguez-Fernandez, M., Mendes, P., Banga, J.R. (2006). A hybrid approach for efficient and robust parameter estimation in biochemical pathways. *Biosystems* **83**:248 – 65.

[105] Bongard, J., Lipson, H. (2007). Automated reverse engineering of nonlinear dynamical systems. *Proc. Natl. Acad. Sci. U.S.A.* **104**:9943 – 8.

[106] Jayawardhana, B., Kell, D.B., Rattray, M. (2008). Bayesian inference of the sites of perturbations in metabolic pathways via Markov Chain Monte Carlo. *Bioinformatics* **24**:1191 – 1197.

[107] Wilkinson, S.J., Benson, N., Kell, D.B. (2008). Proximate parameter tuning for biochemical networks with uncertain kinetic parameters. *Mol. Biosyst.* **4**:74 – 97.

[108] Bäck, T., Fogel, D.B., Michalewicz, Z. (1997). *Handbook of evolutionary computation*. IOPPublishing/Oxford University Press, Oxford.

[109] Corne, D., Dorigo, M., & Glover, F. (1999). *New ideas in optimization*. McGraw Hill, London.

[110] Kell, D.B., Darby, R.M., Draper, J. (2001). Genomic computing: explanatory analysis of plant expression profiling data using machine learning. *Plant Physiol.* **126**:943 – 951.

[111] Salter, G.J., Kell, D.B. (1995). Solvent selection for whole cell biotransformations in organic media. *CRC Crit. Rev. Biotechnol.* **15**:139 – 177.

[112] Shuker, S.B., Hajduk, P.J., Meadows, R.P., Fesik, S.W. (1996). Discovering high-affinity ligands for proteins: SAR by NMR. *Science* **274**:1531 – 4.

[113] Schneider, G., Lee, M.L., Stahl, M., Schneider, P. (2000). *De novo* design of molecular architectures by evolutionary assembly of drug-derived building blocks. *J. Comput. Aided Mol. Des.* **14**:487 – 94.

[114] Rees, D.C., Congreve, M., Murray, C.W., Carr, R. (2004). Fragment-based lead discovery. *Nat. Rev. Drug Discov.* **3**:660 – 72.

[115] Erlanson, D.A., Hansen, S.K. (2004). Making drugs on proteins: site-directed ligand discovery for fragment-based lead assembly. *Curr. Opin. Chem. Biol.* **8**:399 – 406.

[116] Erlanson, D.A., McDowell, R.S., O'Brien, T. (2004). Fragment-based drug discovery. *J. Med. Chem.* **47**:3463 – 82.

[117] Hajduk, P.J. (2006). Fragment-based drug design: how big is too big? *J. Med. Chem.* **49**:6972 – 6.

[118] Leach, A.R., Hann, M.M., Burrows, J.N., Griffen, E.J. (2006). Fragment screening: an introduction. *Mol. Biosyst.* **2**:430 – 46.

[119] Albert, J.S., Blomberg, N., Breeze, A.L., Brown, A.J., Burrows, J.N., Edwards, P.D., Folmer, R.H., Geschwindner, S., Griffen, E.J., Kenny, P.W., Nowak, T., Olsson, L.L., Sanganee, H., Shapiro, A.B. (2007). An integrated approach to fragment-based lead generation: philosophy, strategy and case studies from AstraZeneca's drug discovery programmes. *Curr. Top. Med. Chem.* **7**:1600 – 29.

[120] Alex, A.A., Flocco, M.M. (2007). Fragment-based drug discovery: what has it achieved so far? *Curr. Top. Med. Chem.* **7**:1544 – 67.

[121] Ciulli, A., Abell, C. (2007). Fragment-based approaches to enzyme inhibition. *Curr. Opin. Biotechnol.* **18**:489 – 96.

[122] Hajduk, P.J., Greer, J. (2007). A decade of fragment-based drug design: strategic advances and lessons learned. *Nat. Rev. Drug Discov.* **6**:211 – 9.

[123] Hubbard, R.E., Davis, B., Chen, I., Drysdale, M.J. (2007). The SeeDs approach: integrating fragments into drug discovery. *Curr. Top. Med. Chem.* **7**:1568 – 81.

[124] Hubbard, R.E., Chen, I., Davis, B. (2007). Informatics and modeling challenges in fragment-based drug discovery. *Curr. Opin. Drug Discov. Devel.* **10**:289 – 97.

[125] Jhoti, H., Cleasby, A., Verdonk, M., Williams, G. (2007). Fragment-based screening using X-ray crystallography and NMR spectroscopy. *Curr. Opin. Chem. Biol.* **11**:485 – 93.

[126] Jhoti, H. (2007). Fragment-based drug discovery using rational design. *Ernst Schering Found Symp. Proc.* 169 – 85.

[127] Morphy, R., Rankovic, Z. (2007). Fragments, network biology and designing multiple ligands. *Drug Discov. Today* **12**:156 – 60.

[128] Siegal, G., Ab, E., Schultz, J. (2007). Integration of fragment screening and library design. *Drug Discov. Today* **12**:1032 – 9.

[129] Hesterkamp, T., Whittaker, M. (2008). Fragment-based activity space: smaller is better. *Curr. Opin. Chem. Biol.* **12**:260 – 8.

[130] Hubbard, R.E. (2008). Fragment approaches in structure-based drug discovery. *J. Synchrotron Res.* **15**:227 – 230.

[131] Dobson, P.D., Patel, Y., Kell, D.B. (2008). "Metabolite-likeness" as a criterion in the design and selection of pharmaceutical drug libraries. *Drug Disc. Today*, **14**(1–2):31 – 40.

[132] Willett, P., Barnard, J.M., Downs, G.M. (1998). Chemical similarity searching. *J. Chem. Inf. Comp. Sci.* **38**:983 – 996.

[133] Bajorath, J. (2004). *Chemoinformatics: concepts, methods and tools for drug discovery.* Humana Press, Totowa, NJ.

[134] Maldonado, A.G., Doucet, J.P., Petitjean, M., Fan, B.T. (2006). Molecular similarity and diversity in chemoinformatics: from theory to applications. *Mol. Divers.* **10**:39 – 79.

[135] Eckert, H., Bajorath, J. (2007). Molecular similarity analysis in virtual screening: foundations, limitations and novel approaches. *Drug Discov. Today* **12**:225 – 33.

[136] Russell, B. (1947). *Am I an atheist or an agnostic?*

[137] Harnad, S., Brody, T., Vallieres, F., Carr, L., Hitchcock, S., Gingras, Y., Oppenheim, C., Hajjem, C., Hilf, E.R. (2008). The access/impact problem and the green and gold roads to open access: An update. *Serials Review* **34**:36 – 40.

[138] Kostoff, R.N. (2002). Overcoming specialization. *Bioscience* **52**:937 – 941.

[139] Kell, D.B., Oliver, S.G. (2004). Here is the evidence, now what is the hypothesis? The complementary roles of inductive and hypothesis-driven science in the post-genomic era. *Bioessays* **26**:99 – 105.

[140] Hoffmann, R., Krallinger, M., Andres, E., Tamames, J., Blaschke, C., Valencia, A. (2005). Text mining for metabolic pathways, signaling cascades, and protein networks. *Sci. STKE* **2005,** pe21.

[141] Ananiadou, S., McNaught, J. (2006). *Text mining in biology and biomedicine.* Artech House, London.

[142] Ananiadou, S., Kell, D.B., Tsujii, J.-I. (2006). Text Mining and its potential applications in Systems Biology. *Trends Biotechnol.* **24**:571 – 579.

[143] Jensen, L.J., Saric, J., Bork, P. (2006). Literature mining for the biologist: from information retrieval to biological discovery. *Nat. Rev. Genet.* **7**:119 – 29.

[144] Torvik, V.I., Smalheiser, N.R. (2007). A quantitative model for linking two disparate sets of articles in MEDLINE. *Bioinformatics* **23**:1658 – 65.

[145] Smalheiser, N.R., Swanson, D.R. (1998). Using ARROWSMITH: a computer-assisted approach to formulating and assessing scientific hypotheses. *Comput. Methods Programs Biomed.* **57**:149 – 53.

[146] Swanson, D.R., Smalheiser, N.R., Torvik, V.I. (2006). Ranking indirect connections in literature-based discovery: The role of medical subject headings. *J. Amer. Soc. Inf. Sci. Technol.* **57**:1427 – 1439.

[147] Weeber, M., Kors, J.A., Mons, B. (2005). Online tools to support literature-based discovery in the life sciences. *Brief Bioinform.* **6**:277 – 86.

[148] van der Eijk, C.C., van Mulligen, E.M., Kors, J.A., Mons, B., an den Berg, J. (2004). Constructing an associative concept space for literature-based discovery. *J. Amer. Soc. Information Sci. Technol.* **55**:436 – 444.

[149] Lindsay, R.K., Gordon, M.D. (1999). Literature-based discovery by lexical statistics. *J. Amer. Soc. Inf. Sci.* **50**:574 – 587.

[150] Yetisgen-Yildiz, M., Pratt, W. (2006). Using statistical and knowledge-based approaches for literature-based discovery. *J. Biomed. Informatics* **39**:600 – 611.

[151] Weeber, M., Kors, J.A., Mons, B. (2005). Online tools to support literature-based discovery in the life sciences. *Briefings in Bioinformatics* **6**:277 – 286.

[152] Kostoff, R.N., Briggs, M.B., Solka, J.L., Rushenberg, R.L. (2008). Literature-related discovery (LRD): Methodology. *Technol. Forecast. Soc. Change*, doi:10.1016/j.techfore.2007.11.010.

THE CHEMISTRY OF SIGNAL TRANSDUCTION IN THE TetR SYSTEM

HARALD LANIG AND TIMOTHY CLARK[*]

Computer-Chemie-Centrum, Friedrich-Alexander Universität Erlangen-Nürnberg, Nägelsbachstraße 25, 90152 Erlangen, Germany

E-Mail: *Tim.Clark@chemie.uni-erlangen.de

Received: 28th October 2008 / Published: 16th March 2009

ABSTRACT

Signal transduction proteins in biological systems must be very flexible to undergo the allosteric changes necessary for their function. It is current practice to investigate the modes of action of these systems by X-ray spectroscopy of the different states trapped as crystals. Unfortunately, the forces acting on the proteins by packing effects may lead to distortions comparable to the changes that occur during the allosteric movements. This makes it questionable as to whether X-ray structures can be used to deduce induction mechanisms. In this work, we show for DNA-binding tetracycline repressor proteins that molecular dynamics simulations offer an interesting alternative for determining the induction state and possible mechanisms switching between them. Based on data sampled for different repressor classes with several force field parameter sets, we show that MD simulations have convincing advantages over the analysis of static structures influenced by crystal packing.

INTRODUCTION

Signal transduction is the mechanism by which biological processes are turned on and off. Organisms use signal-transduction pathways to control their development, react to external stimuli such as heat, cold, excess or lack of nutrients, toxic chemicals etc., but also to regulate cell growth and death. Malfunction of signal-transduction systems can lead, among others, to cancer or autoimmune diseases. The effects of many hereditary diseases can also be attributed to malfunction in signal-transduction processes. Thus, signal-transduction pathways and processes play a major role in the systems chemistry of living systems because they combine to form complex signaling networks that control the development, metabolism and defense mechanisms of living organs.

Perhaps the structurally and mechanistically best characterized signal-transduction protein is the tetracycline repressor (TetR)[1]. TetR switches an efflux-pump defense mechanism [2] in resistant Gram-negative bacteria when they are subjected to tetracycline antibiotics.

Tetracycline (Tc)   5a,6-Anhydrotetracycline (ATc)

The parent compound tetracycline (Tc, shown above) is a natural product produced by several *Streptomyces* strains. The metabolite 5a,6-anhydrotetracycline (ATc, show above) is not an effective antibiotic but induces TetR approximately 500 times better than Tc [3] and is therefore often used in biological experiments rather than a tetracycline antibiotic.

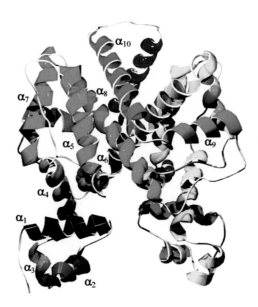

Figure 1. Schematic view of the tetracycline repressor as its homodimer. The α-helices of the front monomer are numbered. The DNA-binding heads contain a helix-turn-helix motif and consist of the first three helices.

Apart from the therapeutic use of tetracycline derivatives (doxycycline is the most commonly used tetracycline and a new derivative, tigecycline, was introduced recently), the importance of the tetracycline/TetR system lies in its use as a "gene switch" [4]. TetR can be introduced into both prokaryotes and eukaryotes in order to be able to turn specific genes on and off at will by administering tetracyclines. TetR has thus become an important tool in molecular biology.

The mode of action and mechanism of induction of TetR are therefore important, both for developing more specific TetR switches that can be used in parallel in one organism and also because of the general significance of signal transduction in biological processes. Quite generally, repressor proteins that switch the expression of other proteins on and off bind to the promoter region of a gene and in doing so prevent expression of the encoded protein by blocking access of RNA-polymerase, the enzyme that synthesizes transfer RNA based on the sequence of the gene. Switching occurs when the repressor protein undergoes an allosteric rearrangement that weakens its binding to the promoter, so that the RNA-polymerase can displace it and proceed to transcribe the gene. In bacteria, TetR binds to the promoters belonging to the genes that encode the tetracycline antiporter (TetA), a membrane-bound protein that pumps tetracycline as the complex of its anion with Mg^{2+} out of the cell. TetR, however, also regulates its own expression, so that when TetA has pumped all the tetracycline out of the cell, more TetR is available to turn off the expression of the two proteins once more. Thus, any disadvantages that the bacterium may suffer from the presence of TetA in its cell wall are avoided by the switching mechanism.

The structure of TetR is shown schematically in Figure 1 and its complex with DNA [5] in Figure 2.

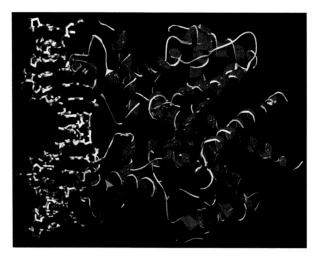

Figure 2. The homodimeric tetracycline repressor bound to DNA (structure taken from PDB entry 1QPI [5])

The contact between the protein and DNA occurs via the helices $\alpha3$, so that the distance between the centers of gravity of the C_α atoms of these two helices (denoted R($\alpha3$-$\alpha3'$) in the following) determines whether the repressor fits into the major groove of the double helix, and therefore how well it binds.

MOLECULAR DYNAMICS SIMULATIONS

The mechanism of induction of TetR was investigated using molecular dynamics (MD) simulations. In this technique, a classical force field (in this case AMBER Parm94 [6], Gromos87 [7] or OPLS_AA [8]) has been used to simulate the movements of the protein by solving Newton's second law of motion starting from random velocities for the atoms [9]. Because the TetR protein is very flexible, implicit water solvent must be used and long-range electrostatic interactions included using the particle-mesh Ewald (PME) technique [10]. The details of the simulations are given in the original article [11].

The main problem in determining the mechanism of induction, which was thought to occur on a timescale of microseconds to milliseconds, was that the length of time that could be simulated when the original studies were carried out was limited to tens of nanoseconds. Thus, it was impossible to observe the allosteric change on induction in the simulations. However, facile rearrangements of biological macromolecules reveal themselves in the normal vibrations. The normal modes with the lowest frequency and the largest amplitude are usually those that lead to the rearrangement. It is not necessary to calculate the force-constant matrix for the system as the normal modes can be calculated by diagonalizing the mass-weighted covariance matrix obtained from an MD simulation [12].

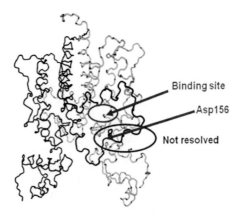

Figure 3. C_α trace of the backbone of TetR showing the flexible loop, which is not resolved in most X-ray structures, the tetracycline binding site and the position of Asp156.

Figure 3 shows the main features of the induction mechanism revealed by the lowest-energy normal mode. The so-called flexible loop, indicated by the blue oval in Figure 3, has only recently been resolved in some X-ray structures. The lowest-energy normal mode revealed a very large movement of aspartate 156 away from the tetracycline binding site towards the DNA-binding heads. This movement is the key to the induction mechanism, which is shown schematically in Figure 4.

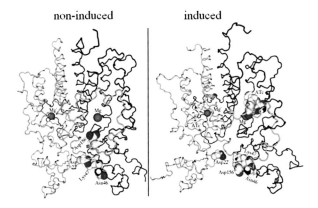

non-induced   induced

Figure 4. Schematic view of the mechanism of induction of TetR by tetracyclines complexed to magnesium. [11] In the non-induced structure (left), Asp156 forms a loose salt bridge to the magnesium ion in the tetracycline binding site and Asp22 forms a tight salt bridge with Lys47 to bind the two DNA-binding heads together. In the induced form (right), ATc displaces Asp156 from the magnesium. The aspartate then migrates towards the DNA-binding heads and forms a new salt bridge with Lys47, thus breaking the one with Asp22. This removes the salt bridge between the two binding heads and allows them to move apart. The flexible loop is highlighted in yellow.

The effect of the inducer (in this case ATc) is to displace Asp156 from the magnesium ion. The aspartate then swings down towards the DNA-binding heads as part of the large movement of the flexible loop observed in the lowest-frequency normal mode. It displaces Lys47 from its salt bridge with Asp22 and thus removes the salt bridge that binds the two DNA-binding heads together.

THE REVERSE PHENOTYPE

One of the most fascinating aspects of TetR is that a single mutation is enough to reverse the behavior of the repressor [13]. The reverse phenotype is produced, for instance, by mutating Gly95 in the wild type TetR class BD to glutamate. We will designate this mutant revTetR. RevTetR is induced in the absence of tetracyclines, but not in their presence. Its induction behavior is thus exactly the opposite of the wild type. We have also performed long MD simulations on this mutant [14] and conclude that the mechanism of induction is exactly the same as that observed for wild type TetR. However, the structure of the revTetR dimer without tetracycline differs significantly from that of the wild type.

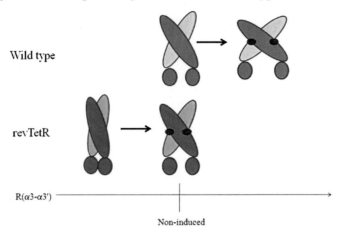

Figure 5. Schematic diagram of the mechanism of induction of the TetR wild type (green above) and the reverse phenotype (blue, below). The horizontal axis represents the distance between the DNA-binding heads, which is optimal for binding to DNA at the distance marked "Non-induced". Docked tetracycline molecules are indicated as red ovals.

Figure 5 shows the situation schematically. The $R(\alpha3-\alpha3')$ distance is too short to bind effectively to DNA in the wild type without tetracycline. The induction movement is the same in the two TetR variants, but in the wild type it extends $R(\alpha3-\alpha3')$ to a value too large to to bind ideally to DNA, whereas in revTetR it increases $R(\alpha3-\alpha3')$ to exactly the value needed for optimal binding. Note, however, that revTetR shows signs of denaturing far more easily than the wild type and is stabilized by tetracyclines [15], so that denaturation (and hence induction) may occur in the absence of tetracyclines. Such an effect is too slow to be revealed by the MD simulations.

DETECTING INDUCTION

Experimental determinations of the induction state of TetR are time-consuming and difficult. It would therefore be useful to be able to determine whether a given TetR variant or mutation is induced in the presence of a given inducer. The $R(\alpha3-\alpha3')$ criterion suggested above may, however, indicate induction in X-ray structures or those taken from MD simulations. In the following, we examine these two possibilities.

DETERMINING INDUCTION FROM X-RAY STRUCTURES

A total of 14 X-ray structures of TetR have been published [16 – 24] with and without a variety of inducers and in one case complexed to DNA. One additional TetR class D structure (2VKE) complexed with Tc and Co^{2+} was published 2007 and shows almost no differences in backbone geometry and ligand position when comparing to 2TRT. Figure 6 shows a plot (in chronological order) of the $R(\alpha3-\alpha3')$ distances obtained from these structures. There are actually 16 distances because the structures 2NS7 and 2NS8 contain two different TetR dimers in the unit cell. The points in the graph are color coded according to whether TetR is induced or not in the combination found in the X-Ray structure.

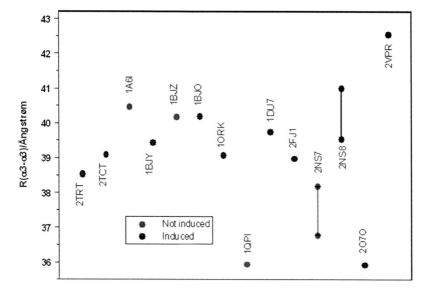

Figure 6. Plot of the $R(\alpha3-\alpha3')$ distances obtained from TetR X-ray structures. The labels for the points indicate the PDB-codes. Two values are given for 2NS7 and 2NS8 because there are two non-equivalent TetR dimers in their unit cells. The points are plotted in the order of their publication.

Figure 6 does not reveal a convincing relationship between $R(\alpha3-\alpha3')$ and the state of induction of TetR. The definitive $R(\alpha3-\alpha3')$ value for the non-induced state of TetR is that for 1QPI, which is the X-ray structure in which the TetR dimer is bound to DNA. However, this distance is almost identical to that found in structure 2O7O, a complex with the strong inducer doxycycline. Quite generally, the $R(\alpha3-\alpha3')$ values from the X-ray structures appear not to be related to the induction state of TetR.

This is not completely surprising. Signal-transduction proteins that switch by undergoing an allosteric change are by their nature very flexible. It is therefore reasonable to expect that the forces needed to switch the conformation are of the same order of magnitude as crystal-packing forces. The structures of signal-transduction proteins in crystals will therefore be perturbed quite strongly from their solution conformations and cannot be expected to reproduce the induction state of the protein correctly. Note, for instance, that if the first five X-ray structures of TetR had been used to propose a mechanism, we would have concluded that the R(α3-α3') distance in the non-induced form is larger than in the induced one – the reverse of what we now believe.

DETERMINING INDUCTION FROM MD SIMULATIONS

Molecular dynamics simulations offer a possible alternative for determining the induction state of TetR. However, for this to be the case, the allosteric rearrangement must take place on a time scale that makes it probable within the length of a simulation, which in our case is 50 – 100 ns. We will show below that, contrary to our expectations, this is the case.

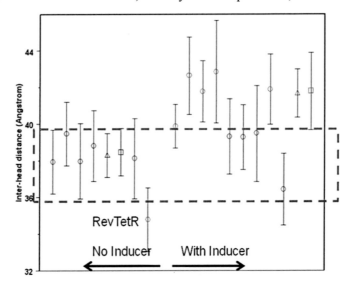

Figure 7. Plot of the R(α3-α3') distances obtained from TetR MD simulations. Red points indicate those that are experimentally induced, blue those that are not. The shapes of the symbols indicate the force field used. Circles are AMBER Parm94, [6] triangles Gromos87 [7] and squares OPLS_AA. [8] The dashed box indicates the limits of R(α3-α3') for non-induced TetR. The error bars represent ± one standard deviation over the sampling period.

Figure 7 shows the results of 19 simulations, all of which were 50 ns or longer. In contrast to the X-ray results, the R(α3-α3') distances from the simulations show a clear relationship to the induction state of TetR. All simulations for which R(α3-α3') falls within the "non-induction window" from approximately 36 – 40 Å represent TetR systems that are found experimentally not to be induced. Those above and below this window are induced.

The differences between some simulations close to the borders are not statistically significant, but nevertheless the simulations provide a remarkably reliable prediction of the induction state.

This result is quite remarkable. MD simulations with three different force fields on a variety of combinations of TetR variants and inducers can reproduce the induction behavior of TetR with far higher reliability than X-ray structures, which are perturbed too much to reflect the induction state correctly. This result even appears to be fairly independent of the force field used. The simulations used three different common force fields, AMBER Parm94, [6] Gromos87 [7] and OPLS_AA [8]. However, many of the simulations did not start in the region of the mean R(α3-α3$'$) values over the sampling periods. The R(α3-α3$'$) values shown in Figure 7 are taken from the later stages (typically after 20 ns or longer) of the simulations after TetR has relaxed to its preferred geometry under the conditions of the simulation. This process in itself is of interest.

ALLOSTERIC REARRANGEMENTS IN MD SIMULATIONS

Figure 8 shows a trace of the R(α3-α3$'$) distance in an AMBER Parm94 simulation of TetR with doxycycline starting from the geometry of the 2O7O X-ray structure. This is the structure that exhibits the same R(α3-α3$'$) distance as 1QPI, the complex with DNA.

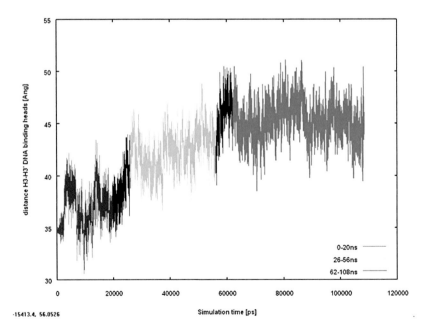

Figure 8. Trace of the R(α3-α3') distance against time for a simulation of TetR with two doxycycline molecules starting from the 2O7O X-ray structure.

The red region of the plot from $0-20$ ns has a mean $R(\alpha3\text{-}\alpha3')$ distance of 36.9 Å, but shows oscillations that appear to increase in amplitude with time. Between 20 and 62 ns, $R(\alpha3\text{-}\alpha3')$ increases steadily to a final value (green region, $62-108$ ns, mean $R(\alpha3\text{-}\alpha3')=45.3$ Å) in which $R(\alpha3\text{-}\alpha3')$ no longer exhibits the strong oscillations seen at the beginning of the simulation. Thus, the simulation remains stable at a non-induced value of $R(\alpha3\text{-}\alpha3')$ for the first 20 ns before undergoing a transition over 40 ns to achieve a final, apparently stable, induced conformation. We emphasize that this process does not necessarily correspond to a real induction event in the biological system, but strictly speaking only to the relaxation of the strained 2O7O X-ray structure. Nevertheless, the fact that the initial conformation remains stable for 20 ns (the complete length of many simulations reported in the current literature) is very significant. This is as far as we are aware the first direct observation in an MD simulation of the pre-equilibrium mechanism of induction [25]. In this mechanism, which has found wide acceptance, the inducer first docks into the non-induced conformation of the repressor to form a metastable complex, which eventually relaxes to the induced conformation of the repressor-inducer complex. In the reverse process, the inducer dissociates from the induced complex to form a metastable, uncomplexed induced conformation of the repressor, which then can relax to the stable non-induced conformation.

CONCLUSIONS

We have shown that MD simulations of $50-100$ ns are an effective tool for identifying the mechanism of induction of TetR and for detecting its state of induction. These conclusions are presumably also valid for other signal-transduction proteins that vary their binding affinity by an allosteric rearrangement. Perhaps predictably, X-ray structures are not well suited for determining the induction state of signal-transduction proteins because we believe the crystal-packing forces to be of the same order of magnitude as those that cause the allosteric rearrangement.

It is perhaps misleading to discuss "the mechanism of induction". Figure 9 illustrates this point.

Thus, not induction is unique, but rather the fact that signal-transduction proteins can bind so strongly to the promoter and that this bound conformation is so robust to environmental factors such as temperature, pH, ion concentrations etc. that should not be able to cause induction. Induction can be caused by just about any structural change, including denaturation, that lowers the binding affinity. Thus, one and the same signal-transduction protein may be induced in many ways. This is in fact true of TetR, which is induced by tetracyclines in the presence of magnesium, by 5a,6-anhydrotetracycline [26] and the TetR-inducing protein TIP [22] in its absence. In the latter case, a different mechanism of induction is observed to that found with tetracycline-magnesium complexes [27].

Figure 9. Schematic illustration of the induction of a signal-transduction protein. The small, specific red area represents the non-induced protein, which can bind strongly to the promoter. The surrounding yellow area represents induced geometries that do not bind as strongly. Clearly, there are very many ways to distort the protein so that it no longer binds strongly.

ACKNOWLEDGMENTS

This work has been supported for the last eight years by the Deutsche Forschungsgemeinschaft as part of SFB 473, *Mechanisms of Transcriptional Regulation*. We thank Wolfgang Hillen, Chris Berens and Peter Gmeiner for many enlightening discussions. None of the results described here would have been obtained without the support of Frank Beierlein, Olaf Othersen, Ute Seidel and Florian Haberl. We gratefully acknowledge large allocations of computer time on HLRB II.

REFERENCES

[1]   Saenger, W., Orth, P., Kisker, C., Hillen, W., Hinrichs, W. (2000) The Tetracycline Repressor-A Paradigm for a Biological Switch. *Angew. Chem. Int. Ed.* **39**:2042 – 2052.

[2]   Walsh, C. (2003) *Antibiotics: Actions, Origins, Resistance.* ASM Press, Washington, D.C.

[3]   Scholz, O., Schubert, P., Kintrup, M., Hillen, W. (2000) Tet Repressor Induction without Mg^{2+}. *Biochemistry* **39**:10914 – 10920.

[4]   Berens, C., Hillen, W. (2003) Gene regulation by tetracyclines. Constraints of resistance regulation in bacteria shape TetR for application in eukaryotes. *FEBS Journal* **270**:3109 – 3121.

[5]   Orth, P., Schnappinger, D., Hillen, W., Saenger, W., Hinrichs, W. (2000) Structural basis of gene regulation by the tetracycline inducible Tet repressor-operator system. *Nature Struct. Biol.* **7**:215 – 219.

[6]   Cornell, W.D., Cieplak, P. Bayly, C. I., Gould, I.R., Merz, K.M. Jr., Ferguson, D.M., Spellmeyer, D.C., Fox, T., Caldwell, J.W., Kollman, P.A. (1995) A Second Generation Force Field for the Simulation of Proteins, Nucleic Acids, and Organic Molecules. *J. Am. Chem. Soc.* **117**:5179 – 5197.

[7]   Van Gunsteren, W.F., Berendsen, H.J.C. (1987) *Groningen Molecular Simulation Library*, Groningen.

[8]   Jorgensen, W.L., Maxwell, D.S., Tirado-Rives, J. (1996) Development and Testing of the OPLS All-Atom Force Field on Conformational Energetics and Properties of Organic Liquids. *J. Am. Chem. Soc.* **118**:11225 – 11236.

[9]   Allen, M.P., Tildesley, D.J. (1987) *Computer Simulation of Liquids*, Oxford University Press, Oxford.

[10]   Darden, T., York, D., Pedersen, L. (1993) Particle mesh Ewald: An N·log(N) method for Ewald sums in large systems. *J. Chem. Phys.* **98**:10089 – 10092.

[11]   Lanig, H., Othersen, O.G., Beierlein, F. R., Seidel, U., Clark, T. (2006) Molecular Dynamics Simulations of the Tetracycline-Repressor Protein: The Mechanism of Induction. *J. Mol. Biol.* **359**:1125 – 1136.

[12]   Amadei, A., Linssen, A.B.M., Berendsen, H.J.C. (1993) Essential Dynamics of Proteins. *Proteins: Struct. Funct. Genet.* **17**:412 – 425.

[13] Scholz, O., Henßler, E.-M., Bail, J., Schubert, P., Bogdanska-Urbaniak, J., Sopp, S., Reich, M., Wisshak, S., Köstner, M., Bertram, R., Hillen, W. (2004) Activity reversal of Tet repressor caused by single amino acid exchanges. *Mol. Microbiol.* **53**:777–789.

[14] Seidel, U., Othersen, O.G., Haberl, F., Lanig, H., Beierlein, F.R., Clark, T. (2007) Molecular Dynamics Characterization of the Structures and Induction Mechanisms of a Reverse Phenotype of the Tetracycline Receptor. *J. Phys. Chem. B* **111**:6006–6014.

[15] Resch, M., Striegl, H., Hennsler, E.M., Sevvana, M., Egerer-Sieber, C., Schiltz, E., Hillen, W., Muller, Y.A. (2008) A protein functional leap: how a single mutation reverses the function of the transcription regulator TetR. *Nucleic Acids Res.* **36**:4390–4401.

[16] Hinrichs, W., Kisker, C., Düvel, M., Müller, A., Tovar, K., Hillen, W., Saenger, W. (1994) Structure of the Tet Repressor-Tetracycline Complex and Regulation of Antibiotic Resistance. *Science* **264**:418–420.

[17] Kisker, C., Hinrichs, W., Tovar, K., Hillen, W., Saenger, W. (1995) The Complex Formed Between Tet Repressor and Tetracycline-Mg^{2+} Reveals Mechanism of Antibiotic Resistance. *J. Mol. Biol.* **247**:260–280.

[18] Orth, P., Cordes, F., Schnappinger, D., Hillen, W., Saenger, W., Hinrichs, W. (1998) Conformational Changes of the Tet Repressor Induced by Tetracycline Trapping. *J. Mol. Biol.* **279**:439–447.

[19] Orth, P., Saenger, W., Hinrichs, W. (1999) Tetracycline-Chelated Mg^{2+} Ion Initiates Helix Unwinding in Tet Repressor Induction. *Biochemistry* **38**:191–198.

[20] Orth, P., Schnappinger, D., Sum, P.-E., Ellestad, G. A., Hillen, W., Saenger, W., Hinrichs, W. (1999) Crystal Structure of the Tet Repressor in Complex with a Novel Tetracycline, 9-(N, N-dimethylglycylamido)-6-demethyl-6-deoxy-tetracycline. *J. Mol. Biol.* **285**:455–461.

[21] Orth, P., Schnappinger, D., Hillen, W., Saenger, W., Hinrichs, W. (2000) Structural basis of gene regulation by the tetracycline inducible Tet repressor-operator system. *Nature Struct. Biol.* **7**:215–219.

[22] Luckner, S.R., Klotzsche, M., Berens, C., Hillen, W., Muller, Y.A. (2007) How an Agonist Peptide Mimics the Antibiotic Tetracycline to Induce Tet-Repressor. *J. Mol. Biol.* **368**:780–790.

[23] Aleksandrov, A., Proft, J., Hinrichs, W., Simonson, T. (2007) Protonation Patterns in Tetracycline:Tet Repressor Recognition: Simulations and Experiments. *ChemBioChem* **8**:675–685.

[24]   Aleksandrov, A., Schuldt, L., Hinrichs, W., Simonson, T. (2008) Tet Repressor Induction by Tetracycline: A Molecular Dynamics, Continuum Electrostatics, and Crystallographic Study. *J. Mol. Biol.* **378**:896–910.

[25]   Kern, D., Zuiderweg, E.R.P. (2003) The role of dynamics in allosteric regulation. *Curr. Opin. Struct. Biol.* **13**:748–757.

[26]   Scholz, O., Schubert, P., Kintrup, M., Hillen, W. (2000) Tet Repressor Induction without Mg^{2+}. *Biochemistry* **39**:10914–10920.

[27]   Haberl, F. (2008) *PhD thesis*, Universität Erlangen-Nürnberg.

PROTEIN INTERACTION, ASSOCIATION AND FIBRILLATION

SARA LINSE

Department of Biophysical Chemistry,
Chemical Center, Lund University,
P O Box 124, S 22100 Lund, Sweden

E-Mail: sara.linse@bpc.lu.se

Received: 6th August 2008 / Published: 16th March 2009

ABSTRACT

A protein can fold efficiently with high fidelity if on average native contacts survive longer than non-native ones. If native contacts survive long enough to obtain a certain level of probability that other native contacts form before the first interacting unit dissociates this provides the folding process with directionality towards the native state and no particular pathway is needed. Interactions among hydrophobic residues are by far more important than electrostatic interactions in protein assembly, folding and stability. Proteins may under certain conditions and as a function of time give up their native folded state and form amyloid fibrils – a process that is involved in a number of human diseases. The fibrillation process can be perturbed by the presence of foreign surfaces, for example nanoparticles of different surface character.

INTRODUCTION

Protein folding and protein folds reflect the large number of non-covalent interactions that form under the very substantial constraints imposed by the covalent chain. Due to steric overlap, roughly ninety percent of the combinations of backbone torsion angles phi and psi are inaccessible. Nevertheless, the accessible ~10% of the Ramachandran map allows for a remarkable variation in protein folds through combinations of extended or helical segments and loops. In this review we will provide examples of studies aiming at understanding the role of fundamental intermolecular interactions in protein folding, binding and the role of foreign surfaces in protein fibrillation.

PROTEIN FOLDING THROUGH KINETIC DISCRIMINATION

Many proteins fold rapidly and spontaneously to the native state. This has puzzled investigators for decades as the process would not be completed within the lifetime of the universe if the protein was deemed to random search through all possible conformations, the Levinthal paradox [1]. Yet, proteins fold on a μs – ms time scale [2 – 5] implying a high degree of directionality of the process. Levinthal interpreted this as an evidence for pathways that direct the search [1] and spontaneous folding is often taken as an evidence that there are one or a few obligatory intermediate structures that the chain must adopt on its way from unfolded to the native state. However, for many proteins no intermediates have been detected and they are classified as two-state folders. The "new view" and funnel model invokes parallel routes for ensembles of proteins [6, 7]. By Monte Carlo simulations we resolved the Levinthal paradox and showed that a protein can fold efficiently with high fidelity if on average native contacts survive longer than non-native ones (Fig. 1) [8]. An important consequence of this finding is that no pathway needs to be specified. Instead, kinetic discrimination among formed contacts is a sufficient criterion for rapid folding to the native state. Successful folding requires that native contacts survive long enough to obtain a certain level of probability that other native contacts form before the first interacting unit dissociates. A modest degree of cooperativity among the native contacts shifts the required ratio of dissociation rates into a realistic regime and makes folding a stochastic process with a nucleation step [8].

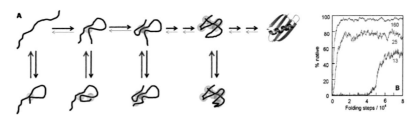

Figure 1. Protein folding through kinetic discrimination. Monte Carlo simulations were performed with the condition that native contacts (green) on average survive longer than non-native ones (red) [8]. (**A**) Successful folding occurs if native contacts survive long enough to obtain a certain level of probability that other native contacts form before the first interacting unit dissociates. (**B**) Results of simulation with different ratios of the average life time for dissociation of native *vs* non-native contacts. The actual life-rime of each contact was picked randomly from an exponentially decaying function with native or non-native lifetime. A modest degree of cooperativity was used among the native contacts so that contacts formed next to preexisting contact get prolonged life time [8].

PROTEIN RECONSTITUTION STUDIES OF THE ROLE OF FUNDAMENTAL INTERACTIONS IN PROTEIN ASSEMBLY

The stability of a protein towards denaturation is often high enough that the protein can tolerate one or more interruptions in the polypeptide chain. After separation, the protein fragments may reassemble spontaneously to regain a complex with a structure and function highly similar to that of the intact native protein [9]. Successful reconstitution has been observed for a large number of proteins including thioredoxin, cytochrome c, ribonuclease, dihydrofolate reductase, calbindins D_{9k} and D_{28k}, troponin C, calmodulin and Trp repressor [10 – 20]. The protein reconstitution process leads to assembly of the fold from more than one chain segment using native contacts in a reaction akin to intramolecular folding.

By the same reasoning, the folding of intact proteins can also be thought of as an association reaction, but one in which the binding partners are imprisoned in the same chain. The analogy between folding and reconstitution has prompted investigators to study the reconstitution reaction as an alternative way to gain insight into protein folding and factors favoring the native states of protein [21 – 27]. In a recent review we made an attempt to summarize the insights gained from reconstitution and folding studies into the molecular factors governing these two processes [28].

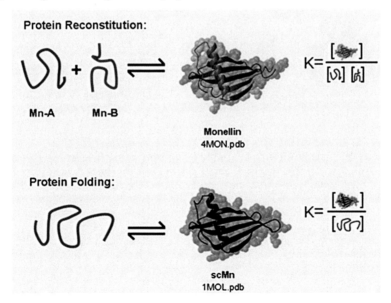

Figure 2. Protein reconstitution and intramolecular folding. The two equilibria for protein reconstitution from two fragments (top) and protein folding of a single covalent chain (bottom) are illustrated for the protein monellin. The equilibrium constants for the two folding processes reflect the difference in their molecularity, with consequent need to define the standard state for the intermolecular reconstitution reaction.

While folding is unimolecular process, protein reconstitution is a bimolecular or higher order process (Figure 2). The fraction of fragments associating as a reconstituted complex will like for all binding reactions be governed by the total concentration of fragments and the equilibrium constant for complex formation, while for the intact protein the fraction of folded protein will be independent of protein concentration (within reasonable limits avoiding aberrant aggregation).

Thereby the reconstitution reactions gain a very significant practical advantage by bringing protein folding under the control of mass action. We have utilized this feature to develop direct experimental approaches to studies of contributions of individual residues to protein assembly (Figure 3; [18, 19, 29, 30]). Our approaches can be used under physiologically relevant solution conditions, and at the conditions under which the protein is maximally stable. This is not possible in traditional protein stability studies which require perturbation of the equilibrium away from the folded state using harsh conditions.

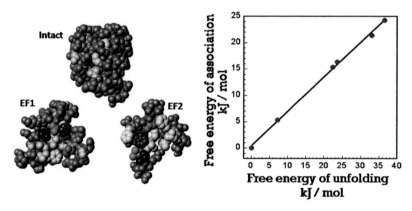

Figure 3. Role of hydrophobic residues in protein assembly and stability. CPK model of intact calbindin D9k (4icb.pdb) and two fragments representing its two EF-hands, EF1 and EF2 (left). The fragments are cut out of the intact structure file, slided apart and rotated 90 degrees towards the viewer to reveal the interaction surface. Hydrophilic backbone and side-chains are colored blue and red, while hydrophobic groups are colored yellow or gold. In this study, the hydrophobic residues with side-chains colored cyan, magenta, green and black were mutated to alter the hydrophobicity and the effects on free energy of fragment association was measured and correlated to the free energy of unfolding of the corresponding intact proteins carrying the same substitutions (right).

The main conclusion from our protein reconstitution studies [18, 19, 29, 30] is that interactions involving the hydrophobic core residues are by far more important for folding and assembly of the protein structure compared to interactions involving charged residues. This settles a long-standing debate as to which physical interactions are most relevant in determining protein tertiary structure.

Three-dimensional domain swapping is a process in which part of the tertiary fold of one chain is replaced by the corresponding part of another chain [17, 31, 32]. Three-dimensional domain swapping may lead to formation of a dimer [33 – 37], or higher multimeric assembly [38 – 40] in which the native fold of the protein is repeated two or more times through association of subdomains from separate chains. Most of the original native contacts are reformed in a reconstituted protein or a domain-swapped oligomer. Swapping thus has in common with fragment reconstitution that the tertiary fold begins in one chain but is completed using segments that originate from another chain. Domain-swapping has been reviewed recently by Liu and Eisenberg [32], and also by Håkansson and Linse [17] who pointed out its relationship to reconstitution. Numerous proteins are found to undergo both processes using similar chain segments.

PROTEIN FIBRILLATION – THE INFLUENCE OF NANOPARTICLES

The native fold of a protein represents a free energy minimum that is strongly sequence dependent. In contrast, amyloid fibrils seem to represent an alternative free energy minimum that has very wide tolerance for protein sequence. These amyloid fibrils have a characteristic cross-beta structure [41] regardless of the native fold or sequence of the parent protein. Fibril formation is documented for so many proteins that it may well be universal, and it has been proposed to reflect universal properties of proteins [42].

Currently, about 30 different proteins and peptides are known to cause human amyloid disease [for reviews see refs. 43 – 47]. These diseases involve self assembly of soluble proteins into large insoluble fibrils through nucleation-dependent assembly, often *via* the formation of oligomeric structures that possess toxic properties [48, 49]. It has been shown that surfaces presented by lipid bilayers, collagen fibres, polysaccharides, and other liquid-air, liquid-solid or liquid-liquid interfaces can have specific and significant effects in promoting amyloid formation [50 – 55]. These observations suggest that interactions with different surfaces could promote protein self-assembly into amyloid fibrils and enhance protein conformational changes associated with other protein misfolding diseases.

While the molecular events behind the processes leading from native to fibrillar states remain elusive, accumulated data from many studies suggest that fibrillation involves a number of intermediate oligomeric states of different association numbers and structures [46]. The use of agents that interfere with these processes and/or allow for the isolation of intermediate species may help elucidate the molecular mechanism of fibril formation. Such strategies have also therapeutic potential for the treatment of neurodegenerative diseases.

We have recently identified co-polymeric nanoparticles as agents that accelerate the fibrillation of β2-microglobulin, β2m. Specifically, we found that the presence of nanoparticles leads to a shortening of the lag phase for nucleation of the fibrillation process. The likely role of the nanoparticles in this process is binding of β2m to the particles, thereby increasing the local concentration and the likelihood of formation of a critical nucleus for fibrillation [56].

These studies are part of an extensive search for potential hazards with nanoparticles. Our premise has been that it is not the nanoparticles *per se* that constitutes the biological risk factor but the nanoparticle with its corona of associated proteins [57]. Nanoparticles in a biological fluid are invariantly coated with proteins and we have developed methodology for mapping which these proteins are and also to study equilibrium affinities and exchange rates for protein-nanoparticle interactions [58–60].

REFERENCES

[1]   Levinthal, C. (1968) Are there pathways for protein folding? *J. Chim. Phys.* **65**:44.

[2]   Anfinsen, C.B., Haber, E., Sela, M., White, F.H. Jr. (1961) The kinetics of formation of native ribonuclease during oxidation of the reduced polypeptide chain. *Proc. Natl. Acad. Sci. U.S.A.* **47**:1309–1314.

[3]   Hagen, S.J., Hofrichter, J., Szabo, A., Eaton, W. (1996) Diffusion-limited contact formation in unfolded cytochrome c: estimating the maximum rate of protein folding. *Proc. Natl. Acad. Sci. U.S.A* **93**:11615–11617.

[4]   Maxwell, K.L., Wildes, D., *et al.* (2006) Protein folding: defining a "standard" set of experimental conditions and a preliminary kinetic data set of two-state proteins. *Protein Sci.* **14**:602–616.

[5]   Fulton, K.F., Devlin, G.L., Jodun, R.A., Silvestri, L., Bottomley, S.P., Fersht, A.R., Buckle, A.M. (2005) PFD: a database for the investigation of protein folding kinetics and stability. *Nucleic Acids Res.* **33**:D279–83.

[6]   Baldwin, R.L. (1995) The nature of protein folding pathways: the classical versus the new view. *J. Biomolec NMR* **5**:103–109.

[7]   Dill, K.A., Chan, H.S. (1997) From Levinthal to pathways to funnels. *Nature Struct. Biol.* **4**:4–9.

[8]   Linse, S., Linse, B. (2007) Protein folding through kinetic discrimination. *J Am Chem Soc* **129**:8481–8486.

[9]   Richards, F.M. (1958) On the enzymatic activity of subtilisin-modified ribonuclease. *Proc. Natl. Acad. Sci. U.S.A.* **44**:162–166.

[10]   Taniuchi, H., Parr, G.R., Juillerat, M.A. (1986) Complementation in folding and fragment exchange. *Methods Enzymol.* **131**:185–217.

[11]   Tasayco, M.L. Carey, J. (1992) Ordered self-assembly of polypeptide fragments to form native-like dimeric trp repressor. *Science* **255**:594 – 597.

[12]   Hantgan, R.R., Taniuchi, H. (1977) Formation of a biologically-active, ordered complex from 2 overlapping fragments of cytochrome-C. *J. Biol. Chem.* **252**:1367 – 1374.

[13]   Holmgren, A. (1972) Thioredoxin-C' – Reconstitution of an active form of *Escherichia coli* thioredoxin from 2 noncovalently linked cyanogen bromide peptide fragments. *FEBS Lett.* **24**:351 – 354.

[14]   Kobayashi, N., Honda, S., Yoshii, H., Uedaira, H., Munekata, E. (1995) Complement assembly of 2 fragments of the Streptococcal protein-G B1 domain in aqueous solution. *FEBS Lett.* **366**:99 – 103.

[15]   Shaw, G.S., Hodges, R.S., Kay, C.M., Sykes, B.D. (1994) Relative stabilities of synthetic peptide homodimeric and heterodimeric troponin-C domains. *Protein Sci.* **3**:1010 – 1019.

[16]   Shuman, C.F., Jiji, R., Akerfeldt, K.S., Linse, S. (2006) Reconstitution of calmodulin from domains and subdomains: Influence of target peptide. *J. Mol. Biol.* **358**:870 – 881.

[17]   Hakansson, M., Linse, S. (2002) Protein reconstitution and 3D domain swapping. *Curr. Protein Pept. Sci.* **3**:629 – 642.

[18]   Berggard, T., Julenius, K., Ogard, A., Drakenberg, T., Linse, S. (2001) Fragment complementation studies of protein stabilization by hydrophobic core residues. *Biochemistry* **40**:1257 – 1264.

[19]   Xue, W.-F., Carey, J., Linse, S. (2004) Multi-method global analysis of thermodynamics and kinetics in reconstitution of monellin. *Proteins* **57**:586 – 595.

[20]   Linse, S., Thulin, E., Gifford, L.K., Radzewsky, D., Hagan, J., Wilk, R.R., Akerfeldt, K.S. (1997) Domain organization of calbindin D28k as determined from the association of six synthetic EF-hand fragments. *Protein Sci.* **6**:2385 – 2396.

[21]   de Prat-Gay, G., Ruiz-Sanz, J., Davis, B., Fersht, A.R. (1994) The structure of the transition state for the association of two fragments of the barley chymotrypsin inhibitor 2 to generate native-like protein: Implications for mechanisms of protein folding. *Proc. Natl. Acad. Sci. U.S.A.* **91**:10943 – 10946.

[22]   Ruiz-Sanz, J., de Prat-Gay, G., Otzen, D.E., Fersht, A.R. (1995) Protein fragments as models for events in protein folding pathways: Protein engineering analysis of the association of 2 complementary fragments of the barley chymotrypsin inhibitor-2. *Biochemistry* **34**:1695 – 1701.

[23] de Prat-Gay, G. (1996) Association of complementary fragments and the elucidation of protein folding pathways. *Protein Eng.* **9**:843 – 847.

[24] Kang, X.S., Carey, J. (1999) Structural organization in peptide fragments of cyto-chrome c by heme binding. *J. Mol. Biol.* **285**:463 – 468.

[25] Dutta, S., Batori, V., Koide, A., Koide, S. (2005) High-affinity fragment complemen-tation of a fibronectin type III domain and its application to stability enhancement. *Protein Sci.* **14**:838 – 2848.

[26] Mitchinson, C., Baldwin, R.L. (1986) The design and production of semisynthetic ribonucleases with increased thermostability by incorporation of S-peptide analogues with enhanced helical stability. *Proteins.* **1**:23 – 33.

[27] Schultz, D.R., Ladbury, J.E., Smith, G.P., Fox, R.O. (1998) Interactions of ribonu-clease S with ligands from random peptide libraries. In *Applications of Calorimetry in the Biological Sciences.* (ed. J.E. and C. Ladbury), pp. 123 – 138. John Wiley and Sons, Ltd, New York.

[28] Carey, J., Lindman, S., Bauer, M.C., Linse, S. (2008) Protein reconstitution and three-dimensional domain swapping. Limits and benefits of covalency. *Protein Science* **16**:2317 – 2333.

[29] Xue, W.F., Szczepankiewicz, O., Bauer, M.C., Thulin, E., Linse, S. (2006) Intra- versus intermolecular interactions in monellin: Contribution of surface charges to protein assembly. *J. Mol. Biol.* **358**:1244 – 1255.

[30] Dell'Orco, D., Xue, W.F., Thulin, E., Linse, S. (2005) Electrostatic contributions to the kinetics and thermodynamics of protein assembly. *Biophys. J.* **88**:1991 – 2002.

[31] Bennett, M.J., Choe, S. Eisenberg, D. (1994) Domain swapping – Entangling alli-ances between proteins. *Proc. Natl. Acad. Sci. U.S.A.* **91**:3127 – 3131.

[32] Liu, Y., Eisenberg, D. (2002) 3D domain swapping: As domains continue to swap. *Protein Sci.* **11**:1285 – 1299.

[33] Hakansson, M., Svensson, A., Fast, J., Linse, S. (2001) An extended hydrophobic core induces EF-hand swapping. *Protein Sci.* **10**:927 – 933.

[34] Ekiel, I., Abrahamson, M. (1996) Folding-related dimerization of human cystatin C. *J. Biol. Chem.* **271**:1314 – 1321.

[35] Liu, Y.S., Hart, P.J., Schlunegger, M.P., Eisenberg, D. (1998) The crystal structure of a 3D domain-swapped dimer of RNase A at a 2.1-angstrom resolution. *Proc. Natl. Acad. Sci. U.S.A.* **95**:3437 – 3442.

[36]   Byeon, I.J.L., Louis, J.M., Gronenborn, A.M. (2003) A protein contortionist: Core mutations of GBI that induce dimerization and domain swapping (vol 333, pg 141, 2003). *J. Mol. Biol.* **334**:605 – 605.

[37]   Janowski, R., Kozak, M., Jankowska, E., Grzonka, Z., Grubb, A., Abrahamson, M., Jaskolski, M. (2001) Human cystatin C, an amyloidogenic protein, dimerizes through three-dimensional domain swapping. *Nat. Struct. Biol.* **8**:316 – 320.

[38]   Lawson, C.L., Benoff, B., Berger, T., Berman, H.M., Carey, J. (2004) E. coli trp repressor forms a domain-swapped array in aqueous alcohol. *Structure* **12**:1099 – 1108.

[39]   Sanders, A., Craven, C.J., Higgins, L.D., Giannini, S., Conroy, M.J., Hounslow, A.M., Waltho, J.P., Staniforth, R.A. (2004) Cystatin forms a tetramer through structural rearrangement of domain-swapped dimers prior to amyloidogenesis. *J. Mol. Biol.* **336**:165 – 178.

[40]   Guo, Z.F., Eisenberg, D. (2006) Runaway domain swapping in amyloid-like fibrils of T7 endonuclease I. *Proc. Natl. Acad. Sci. U.S.A.* **103**:8042 – 8047.

[41]   Nelson, R., Sawaya, M.R., Balbirnie, M., Madsen, A.O., Riekel, C., Grothe, R., Eisenberg, D. (2005) Structure of the cross-beta spine of amyloid-like fibrils. *Nature* **435**:773 – 778.

[42]   Dobson, C.M. (1999) Protein misfolding, evolution and disease. *Trends Biochem. Sci.* **24**:329 – 332.

[43]   Koo, E.H., Lansbury, P.T. Jr., Kelly, J.W. (1999). Amyloid diseases: abnormal protein aggregation in neurodegeneration. *Proc. Natl. Acad. Sci. U.S.A.* **96**:9989 – 9990.

[44]   Chien, P., Weissman, J.S., DePace, A.H. (2004) Emerging principles of conformation-based prion inheritance. *Ann. Rev. Biochem.* **73**:617 – 656.

[45]   Westermark, P., Benson, M.D., Buxbaum, J.N., Cohen, A.S., Frangione, B., Ikeda, S., Masters, C.L., Merlini, G., Saraiva, M.J., Sipe, J.D. (2005) Amyloid: toward terminology clarification. Report from the Nomenclature Committee of the International Society of Amyloidosis. *Amyloid* **12**:1 – 4.

[46]   Chiti, F., Dobson, C.M. (2006) Protein misfolding, functional amyloid, and human disease. *Ann. Rev. Biochem.* **75**:333 – 366.

[47]   Huff, M.E., Balch, W.E., Kelly, J.W. (2003) Pathological and functional amyloid formation orchestrated by the secretory pathway. *Curr. Opin. Struct. Biol.* **13**:674 – 682.

[48] Cleary, J.P., Walsh, D.M., Hofmeister, J.J., Shankar, G.M., Kuskowski, M.A., Selkoe, D.J., Ashe, K.H. (2005) Natural oligomers of the amyloid-beta protein specifically disrupt cognitive function. *Nat. Neurosc.* **8**:79–84.

[49] Baglioni, S., Casamenti, F., Bucciantini, M., Luheshi, L.M., Taddei, N., Chiti, F., Dobson, C.M., Stefani, M. (2006) Prefibrillar amyloid aggregates could be generic toxins in higher organisms. *J. Neurosci.* **26**:8160–8167.

[50] Knight, J.D., Miranker, A.D. (2004) Phospholipid catalysis of diabetic amyloid assembly. *J. Mol. Biol.* **341**:1175–1187.

[51] Relini, A., Canale, C., De Stefano, S., Rolandi, R., Giorgetti, S., Stoppini, M., Rossi, A., Fogolari, F., Corazza, A., Esposito, G. *et al.* (2006). Collagen plays an active role in the aggregation of beta2-microglobulin under physiopathological conditions of dialysis-related amyloidosis. *J. Biol. Chem.* **281**:16521–16529.

[52] Yamaguchi, I., Suda, H., Tsuzuike, N., Seto, K., Seki, M., Yamaguchi, Y., Hasegawa, K., Takahashi, N., Yamamoto, S., Gejyo, F. *et al.* (2003). Glycosaminoglycan and proteoglycan inhibit the depolymerization of beta2-microglobulin amyloid fibrils *in vitro. Kidney Int.* **64**:1080–1088.

[53] Myers, S.L., Jones, S., Jahn, T.R., Morten, I.J., Tennent, G.A., Hewitt, E.W., Radford, S.E. (2006) A systematic study of the effect of physiological factors on beta2-microglobulin amyloid formation at neutral pH. *Biochemistry* **45**:2311–2321.

[54] Powers, E.T., Kelly, J.W. (2001) Medium-dependent self-assembly of an amphiphilic peptide: direct observation of peptide phase domains at the air-water interface. *J. Am. Chem. Soc.* **123**:775–776.

[55] Lu, J.R., Perumal, S., Powers, E.T., Kelly, J.W., Webster, J.R., Penfold, J. (2003) Adsorption of beta-hairpin peptides on the surface of water: a neutron reflection study. *J. Am. Chem. Soc.* **125**:3751–3757.

[56] Linse, S., Cabaliero-Lago, C., Xue, W.-F., Lynch, I., Lindman, S., Thulin, E., Radford, S.R., Dawson, K.A. (2007) Nucleation of protein fibrillation by nanoparticles. *Proc. Natl. Acad. Sci. U.S.A.* **104**:8691–8696.

[57] Lynch, I., Dawson, K.A., Linse, S. (2006) Detecting cryptic epitopes created by nanoparticles. *Science STKE* **327**, pp. pe14.

[58] Cedervall, T., Lynch, I., Lindman, S., Berggård, T., Thulin, E., Nilsson, H., Dawson, K.A., Linse, S. (2007) Understanding the nanoparticle-protein corona using methods to quantify exchange rates and affinities of proteins for nanoparticles. *Proc. Natl. Acad. Sci. U.S.A.* **104**:2050–2055.

[59]   Cedervall, T., Lynch, I., Foy, M., Berggård, T., James, P., Donnelly, S.C., Cagney, G., Linse, S., Dawson, K.A. (2007) Detailed Identification of Plasma Proteins Absorbed to Copolymer Nanoparticles. *Angewandte Chemie* **46**:5754 – 5756.

[60]   Lindman, S., Lynch, I., Thulin, E., Nilsson, H., Dawson, K.A., Linse, S. (2007) Systematic invesitigation of the thermodynamics of HSA adsorption to N-iso-propyl-acrylamide N-tert-butylacrylamide copolymer nanoparticles. Effects of particle size and hydrophobicity. *Nanoletters* **7**:914 – 920.

Beilstein-Institut   Systems Chemistry, May 26th – 30th, 2008, Bozen, Italy

SHEDDING LIGHT ON NUCLEIC ACIDS AND DNA UNDER CONSTRUCTION

ALEXANDER HECKEL

University of Frankfurt,
Cluster of Excellence Macromolecular Complexes,
Max-von-Laue-Str. 9, 60438 Frankfurt am Main, Germany

E-Mail: heckel@uni-frankfurt.de

Received: 4th June 2008 / Published: 16th March 2009

ABSTRACT

The first part of our research deals with the spatiotemporal regulation of biological processes. We use light as addressing mechanism and modify nucleic acids in a way to make them light-responsive. Light has the advantage that it can be easily generated and manipulated with well-established (laser and microscope) technologies and many of the currently investigated model organisms are light-accessible. On the other hand nucleic acids are the base of powerful techniques such as for example RNA interference for the regulation of genes and aptamers for the regulation of the function of proteins.

In the second part of our research we are exploring new ways to assemble nanometer-scaled objects from DNA but instead of only relying on the Watson-Crick interaction we are using alternative – orthogonal – interaction strategies like for example "Dervan-type poly-amides" which can sequence-selectively bind to double-stranded DNA. Two of these polyamides – combined via a linker – form a "DNA strut" which can sequence-selectively "glue" together two DNA double helices.

Light-activatable ("Caged") Nucleic Acids – Photochemistry in Living Cells

Most of the processes in living organisms are exquisitely spatiotemporally regulated – and this is true at every level of organization. A cell is more than just the sum of its individual (non-interacting) constituents, a tissue is more than just an assembly of individual cells and an organism is more than just a statistic assembly of tissues. This aspect of complicated but well-choreographed systems is for example especially important in developmental biology or neurology. If one wants to understand the systems involved or diseases resulting from perturbations thereof, a prerequisite is that we can ask nature our questions in equally spatiotemporally well-defined manners – hence the need for tools that allow us to set biological stimuli with exact control of the location, the point in time and the magnitude. This requirement calls for a broadly applicable addressing mechanism and light is ideally suited in this respect. Light is an orthogonal trigger signal because only a minority of biological systems is already light-responsive by themselves. Light is also a "harmless" trigger signal if one chooses the right wavelength. Additionally many model organism or even tissues are light-accessible and the technologies to generate and manipulate light are very well established. For example a confocal microscope (with appropriate lasers) contains already everything one needs: The laser light sources, the scanning system with mirrors to position the beam in arbitrarily regions of the sample and the confocal setup to detect what is happening upon stimulation. Using two-photon excitation it is even possible to irradiate three-dimensionally well-defined volume elements – as if a "light cursor" were moving through the object of interest [1 – 2]. These are the reasons why we got interested in doing "photochemistry in living cells".

The question now arises how to couple a trigger signal such as light to a biological process so that for example genes or proteins can be switched on or off. There are several ways to realize this. All of them involve either using photolabile "protecting" groups or bi- or multistable photoswitchable systems [3]. The latter approach of attaching "photoswitches" to biologically active molecules has the appealing benefit of reversible switchability but it is certainly very difficult to realize. In this case both the ON- and the OFF-state are modified compounds and this makes it difficult to obtain a clear "binary" switching behavior which is needed for most of the subsequent experiments. The approach of attaching a photolabile group to the active site of a biologically active substance is much easier to realize. The price to pay is that these compounds can only be irreversibly activated but in many cases this is already enough. The general concept of temporarily blocking the biological activity of a compound has already been realized thirty years ago when Hoffman at Yale prepared a derivative of ATP which he called "caged ATP" [4]. Caged ATP has up to now been used in a vast number of studies and is also commercially available nowadays. On the other hand only very few studies existed with caged nucleic acids until some years ago. This is how we got interested in this field.

Our approach of making nucleic acids light-responsive is to attach photolabile cageing groups to the nucleobases (Fig. 1). These cages act both as steric block and also perturb the hydrogen bonding capabilities of the nucleobases. Hence they can be seen as temporary mismatches. This can for example be demonstrated by melting temperature studies. Before photolysis the melting temperature of a duplex is reduced significantly if caged residues are introduced whereas after photolysis the melting temperature of an unmodified duplex is recovered [5].

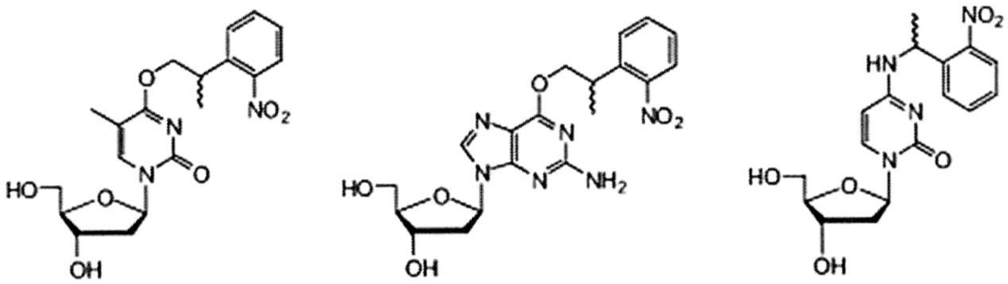

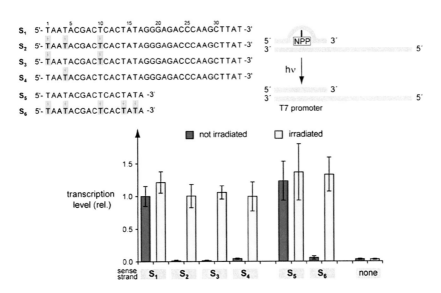

Figure 1. Examples of our nucleobase-caged residues for light-activatable DNAs and RNAs. The photolabile NPP group had been introduced by Pfleiderer *et al.* [6] and is an interesting alternative to the NPE group. Instead of generating a nitrosocompound upon photolysis this group yields α-methylnitrostyrene.

Figure 2. An example for light-induced transcription. Caged residues act as local perturbations in a T7 promoter and prohibit transcription until the photolysis event in which they are fully removed [5].

In a first attempt to demonstrate the usefulness of this type of caged nucleic acids we began to study light-induced transcription (Fig. 2). Therefore we used a luciferase gene which was under the control of a T7 promoter. Normally the T7 RNA polymerase would recognize this promoter and start transcription. However, caged residues in the double-stranded promoter region should result in a local perturbation which prevents the T7 RNA polymerase from recognizing its target binding site and hence transcription should not occur until – after irradiation – the perturbation is removed and the unmodified double-stranded promoter is regenerated. Figure 2 shows the results we obtained: It can be seen that no matter where the cage was introduced in the sense strand no transcription could be observed before irradiation within error limits. However, after irradiation in every case as much transcript was produced as if there had never been any modification. In particular one cage was enough to completely prohibit transcription but even five cages could still be completely photolyzed.

We then proceeded to check if it was also possible to put RNA interference under the control of a light trigger signal. The central players in RNAi are the siRNAs. However, upon introduction of mismatches siRNAs become miRNAs which act in a different mechanism but are certainly still active. However, it is known that there are regions in which mismatches in siRNAs or miRNAs are not tolerated [7]. Understandably, so this is true for the central region in the so-called guide strand of the double-stranded siRNA which is the one that is incorporated into the RISC complex and guides it as to which mRNA to cut. This point of scission of the mRNA is exactly opposite of the center of the guide strand. Interestingly the same study showed that at this position deoxyresidues are well tolerated. Even though we had already prepared caged RNA it turned out not to be necessary in this case. Thus, we prepared hybrids of siRNAs with one caged deoxyresidue in the middle and thus circumvented several synthetic steps addressing the regiochemical aspects of RNA phosphoramidite synthesis [8]. As can be seen in Figure 3 our caged siRNAs were completely inactive within error limits before irradiation but the activity could be restored very well upon photolysis. In this assay HeLa cells were used which were transiently transfected with plasmids coding for EGFP and RFP. By its sequence siRNA only targeted EGFP expression. It is important to note that the reduction of fluorescence upon photolysis is not due to photobleaching as can be seen from the bars corresponding to the experiments in which (caged) nonsense RNA (nsRNA) targeting neither EGFP nor RFP was used with or without irradiation.

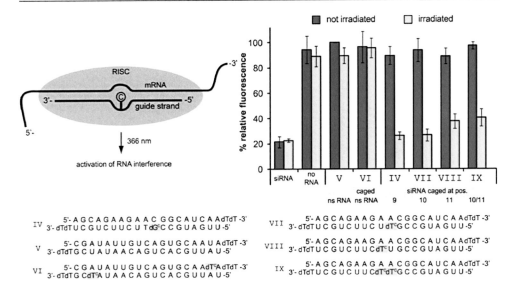

Figure 3. Overview of RNA interference experiments using caged siRNAs [8].

Another interesting question is whether it is not only possible to regulate gene expression by light but also to regulate protein activity – using the same approach based upon caged nucleic acids! At first this might seem like a contradiction but the link is the aptamers technology. Aptamers are single-stranded DNAs or RNAs which can be obtained in a selection process (that does not involve rational design). Using this "SELEX" process it is possible to generate aptamers against very many molecular targets or supramolecular target structures up to even cells [9]. One such aptamers is for example a 15mer DNA sequence which can fold into a G-quadruplex structure and bind and inhibit thrombin which acts as factor IIa of the blood clotting cascade (Fig. 4) [10]. Thus, this aptamers can act as an anticoagulant. From a crystal structure it was known how this aptamers interacts with thrombin (shown as cartoon representation of the PDB structure 1HAO in the upper left corner of Figure 4). This crystal structure identifies four T residues as mediating the inter-action. Thus our rationale was that we block the access to one of them and thus make the aptamers temporarily inactive [11]. Figure 4 also shows the result of blood clotting assays using the caged aptamers variants and the wild type version. As can be seen the caged aptamers variant A_2 was indeed completely inactive and its activity could be regenerated by irradiation – albeit not to 100% in this case. We could show that this is due to a pH-dependent side reaction which is described in a mechanistic paper on the NPP group [6]. This reaction generates a photostable byproduct which is still modified on the nucleobase. Even though this is undesired using a higher amount of caged aptamers this problem can easily be circumvented.

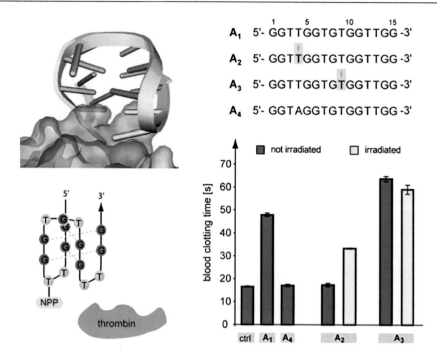

Figure 4. Overview of a DNA aptamers which binds to thrombin (cartoon representation of the PDB crystal structure 1HAO shown in the upper left corner) and a strategy to sterically block its activity with a cageing group [11].

While the previous example shows how aptamers can be switched on with light, the strategy is not yet generally applicable because detailed information about the aptamers-target interaction must be present. However, while it is not easy to obtain for example crystal structures of protein-nucleic acid complexes it is much easier to predict secondary structure formation of DNA or RNA. Since aptamers are only active if they are present in the required conformation a more general approach is to use the cageing technology to trigger the formation of this active conformation with light. The predominant structural feature of the above-mentioned aptamers is the G-quadruplex. Hence, we got interested in the light-triggered formation of G-quadruplex structures. Therefore we chose again the previous aptamers but also a sequence that folds into a three-layer G-quadruplex structure (Fig. 5) [12]. This sequence is derived from human telomers. The presence or absence of the G-quadruplex structure was assayed by CD spectroscopy. Figure 5 shows the CD spectra of the wild type sequence T_1 both in buffer and in ion-free water. The first shows the presence of the G-quadruplex. In the second case no central ions which are required for the formation of the G-quadruplex are present so that the corresponding spectrum can serve as a positive control for the absence of the G-quadruplex. As can be seen the caged sequence T_2 yields the same spectrum even in the presence of buffer salts. Upon irradiation the secondary structure formation is triggered and the same spectrum as the one of the wild type is obtained.

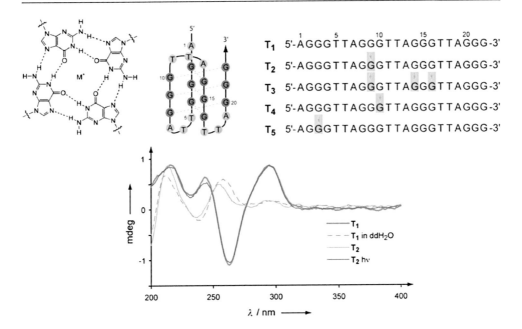

Figure 5. Light-triggered formation of nucleic acid secondary structures [12]. CD spectroscopy shows that only upon irradiation the sequence T_2 forms a G-quadruplex structure.

While the two above examples have shown that it is possible to turn aptamers activity ON with light it might also be desirable to turn aptamers activity OFF. This should also be possible: By attaching an antisense sequence for example to the 5'-end of the anti-thrombin aptamers (cf. Figure 4) the aptamers will not be present in its active conformation any more but rather exist as an inactive hairpin. If the antisense sequence is caged, however, this will not happen until the moment of irradiation (Fig. 6) [13]. It turned out that in addition to the required turn it was sufficient to use only four antisense residues to completely prevent any anticoagulatory activity. As can be seen in Figure 6 aptamer A_4 with the caged antisense-region was active before irradiation but could be efficiently switched off upon irradiation.

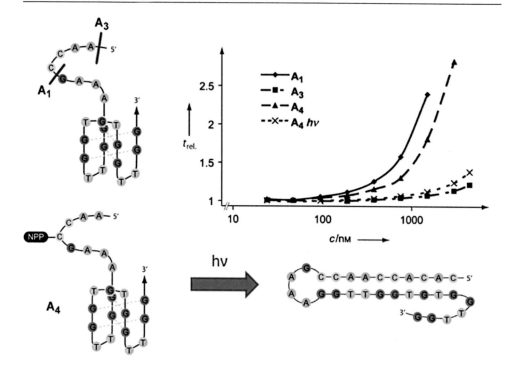

Figure 6. A strategy to turn aptamers OFF with light using a caged antisense-strand. The graph shows dose-dependent anticoagulatory behavior [13].

The above examples demonstrate the broad applicability of the concept of using nucleobase-caged nucleic acids. Ongoing research on all levels of the approach will even improve its versatility and make it suitable for the next step which will be addressing spatiotemporal questions in entire biological systems with all their inherent complexity.

Using DNA as Building Material

The second part of my group addresses an entirely different question. Throughout the last decades computers have become tremendously more powerful. This development has profound implications on society. Examples for these implications can be found everywhere and include among others also the broad availability of the internet which has indeed changed our everyday lives. Still ten years ago mobile phones were far from being common. The steep increase in computing power has also left its footprints in the world of science and made things possible that were still unthinkable several years ago. However, this miniaturization trend cannot go on indefinitely. The reason is that photolithography – the process used for generating the computers' hearts – is using light to transfer the information of a mask onto the silicon surface and even with ideal optics the size of the smallest feature that can be generated is about the wavelength of the light that is used.

This has already been realized over a century ago by Ernst Abbé. Reducing this wavelength indefinitely is also no viable solution. But instead of constantly trying to make things smaller in a "top-down" approach an alternative is to try and use the power of molecular self-assembly ("bottom-up" approach) to generate structures beyond the diffraction limit. This idea is not new and Richard Feynman was one of the most prominent advocates for this idea for example when he gave his famous lecture entitles "There's plenty of room at the bottom" about 50 years ago [14].

Proofs that nano-machineries are possible exist amply inside of living organisms and cells. The ATP synthase or multidrug efflux pumps are beautiful examples of efficient nano-machineries. However, while we have come very far in understanding how these nanoscopic miracles work we are still far from being able to construct similar objects from scratch. This is for example due to the fact that the prediction of protein folding is still very difficult. On the other hand proteins are not the only material from which functional or structural nano-scopic objects can be constructed. Nucleic acids offer many advantages here: They form well-known structures, have predictable, "programmable" interactions (via the Watson-Crick base pairings), can be synthesized in automated fashion and manipulated with estab-lished protocols and have exactly the right dimensions in the nanometer range.

While the usefulness of DNA and RNA as nano-material has already been nicely demon-strated [15] there remains a problem; nucleic acids are linear, "soft" and do not form significant tertiary interactions. The latter is what makes them different from proteins and peptides which offer a richness of possible interactions. To compensate for this we have decided to add other interaction principles to the world of nucleic acid nano-architectures. What this means is that we want to add orthogonal structural elements – orthogonal to the Watson-Crick base pairing.

An example of such orthogonal structural elements – but by far not the only solution – are Dervan-polyamides [16]. This is a set of compounds which binds and sequence-selectively recognizes the DNA minor groove. Dervan-polyamides can be generated to bind to almost any possible DNA sequence. In the simplest realization of our approach to shape the tertiary structure of DNA we have therefore combined two of these polyamides with a flexible linker and constructed what we called a "DNA strut" (Fig. 7) [17]. Such a strut can bind to double-stranded DNA with both its ends and hence hold them together. For this reason one could think of this system as "sequence-selective glue" for DNA-architectures. This new structural element will allow whole new ways of assembling DNA objects.

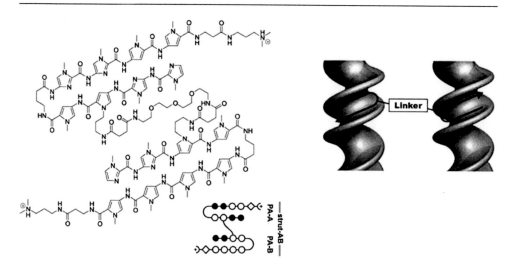

Figure 7. Joining two Dervan-type polyamides (PA-A and PA-B, also shown as cartoon representation) via a linker a DNA strut is obtained which is capable of holding ("gluing") two DNA double strands sequence-selectively together [17].

To prove and quantify that the expected ternary complex between two (different) DNA double helices and the DNA strut is formed we had to explore new techniques because none of the established ones for the characterization of Dervan-polyamides was suitable. Using fluorescence cross-correlation spectroscopy (FCCS) [18] and small DNA hairpins we were able to prove that the expected ternary complex is indeed formed and under the given conditions a dissociation constant of 20 nM was measured (for more details see the original paper [17]). What is equally important is that upon a slight change in DNA sequence no binding could be observed any more.

The next question was to see if DNA objects bigger than hairpins could also be held together using the DNA strut. Therefore, we started constructing small double-stranded DNA rings containing 168 base pairs. These rings are relatively small and shape-persistent compared to common DNA plasmids. The rings could be obtained using repetitive A-tract sequences which are known to be inherently bent. Six constituting oligodeoxynucleotides were ligated and objects with incomplete ring closure were digested using an exonuclease. Again by FCCS we studied whether the ternary complex between two DNA rings and the strut was formed and obtained 30 nM as dissociation constant. This is already a complex of about 208 kDa to which the strut only contributes with about 1%. Interestingly, after replacing three base pairs in the binding site of one polyamide no interaction could be detected any more by FCCS. Encouraged by these results we will now proceed and use this strategy to assemble more complicated objects from nucleic acids as well as new alternative orthogonal approaches for the interaction with double-stranded DNA.

REFERENCES

[1] Zipfel, W.R., Williams, R.M., Webb, W.W. (2003) Nonlinear magic: multiphoton microscopy in the biosciences. *Nat. Biotechnol.* **21**:1369 – 1378.

[2] LaFratta, C.N., Fourkas, J.T., Baldacchini, T., Farrer, R.A. (2007) Multiphoton Fabrication. *Angew. Chem. Int. Ed.* **46**:6238 – 6258.

[3] Mayer, G., Heckel, A. (2006) Biologically active molecules with a "light switch". *Angew. Chem. Int. Ed.* **45**:4900 – 4921.

[4] Kaplan, J.H., Forbush III, B., Hoffman, J.F. (1978) Rapid Photolytic release of adenosine 5'-triphosphate from a protected analogue: utilization by the Na:K pump of human red blood cell chosts. *Biochemistry* **17**:1929 – 1935.

[5] Kröck, L., Heckel, A. (2005) Photoinduced transcription by using temporarily mismatched caged oligonucleotides. *Angew. Chem. Int. Ed.* **44**:471 – 473.

[6] Walbert, S., Pfleiderer, W., Steiner, U.E. (2001) Photolabile protecting groups for nucleosides: mechanistic studies of the 2-(2-nitrophenyl)ethyl group. *Helv. Chim. Acta* **84**:1601 – 1611.

[7] Chiu, Y.L., Rana, T.M. (2003) siRNA function in RNAi: a chemical modification analysis. *RNA* **9**:1034 – 1048.

[8] Mikat, V., Heckel, A. (2007) Light-dependent RNA interference with nucleobase-caged siRNAs. *RNA* **13**:2341 – 2347.

[9] Famulok, M., Hartin, J.S., Mayer, G. (2007) Functional aptamers and aptazymes in biotechnology, diagnosis, and therapy. *Chem. Rev.* **107**:3715 – 3743.

[10] Bock, L.C., Griffin, L.C. Latham, J.A. Vermaas, E.H., Toole, J.J. (1992) Selection of single-stranded DNA molecules that bind and inhibit thrombin. *Nature* **355**:564 – 566.

[11] Heckel, A., Mayer, G. (2005) Light regulation of aptamer activity: an anti-thrombin aptamer with caged thymidine nucleobases. *J. Am. Chem. Soc.* **127**:822 – 823.

[12] Mayer, G., Kröck, L., Mikat, V., Engeser, M., Heckel, A. (2005) Light-induced formation of G-quadruplex DNA secondary structures. *ChemBioChem* **6**:1966 – 1970.

[13] Heckel, A., Buff, M.C.R., Raddatz, M.S.L., Müller, J., Pötzsch, B., Mayer, G. (2006) An anticoagulant with light-triggered antidote activity. *Angew. Chem. Int. Ed.* **45**:6748 – 6750.

[14] http://www.its.caltech.edu/~feynman/plenty.html

[15]   Heckel, A., Famulok, M. (2008) Building objects from nucleic acids for a nanometer world. *Biochimie* **90**(7):1096 – 1107.

[16]   Dervan, P.B. (2001) Molecular recognition of DNA by small molecules. *Bioorg. Med. Chem.* **9**:2215 – 2235.

[17]   Schmidt, T.L., Nandi, C.K., Rasched, G., Parui, P.P., Brutschy, B., Famulok, M., Heckel, A. (2007) Polyamide struts for DNA architectures. *Angew. Chem. Int. Ed.* **46**:4382 – 4384.

[18]   Haustein, E., Schwille, P. (2007). Fluorescence correlation spectroscopy: novel variations of an established technique. *Annu. Rev. Biophys. Biomol. Struct.* **36**:151 – 169.

 **Beilstein-Institut**

Systems Chemistry, May 26th – 30th, 2008, Bozen, Italy

HIGH-THROUGHPUT ANALYSIS OF NUCLEOSIDE- AND NUCLEOTIDE-BINDING BY PROTEINS

JUSTIN K.M. ROBERTS[1*], CECELIA WEBSTER[1], THOMAS C. TERWILLIGER[2] AND CHANG-YUB KIM[2]

[1]Department of Biochemistry, University of California, Riverside, CA 92521, USA.

[2]Bioscience Division, MS M888, Los Alamos National Laboratory, Los Alamos, NM 87545, USA.

E-Mail: *justin.roberts@ucr.edu

Received: 25th June 2008 / Published: 16th March 2009

ABSTRACT

Many proteins function via selective binding of small molecules, and an important class of ligands is nucleosides, including derivative mono- and dinucleotides, which participate in processes such as catalysis and signal transduction. Determining the specificity of nucleoside/nucleotide binding is therefore central to understanding the function of many proteins. We describe use of dye-ligand affinity chromatography methods to identify putative nucleotide-binding proteins, and to determine the specificity of binding to structurally related ligands. In one approach, putative nucleoside-binding proteins are captured from crude protein extracts of cells, for identification via standard proteomic methods. In a second approach, interactions of different nucleosides with purified recombinant protein targets, immobilized on dye, are determined to assess the specificity of ligand recognition by a given nucleotide-binding protein in the context of structural and functional genomics, and drug development.

INTRODUCTION

Nucleosides and their derivatives are molecular carriers central to transmission of energy, genetic information and intracellular signals, serving in these roles usually via specific interactions with proteins [1]. Furthermore, many drugs contain nucleoside-related moieties; for example nucleotide-protein interactions are considered critical in the action of isoniazid, one of the most efficient drugs for the treatment of *Mycobacterium tuberculosis* (Mtb) infections [2] and purine nucleoside analogs have been recently developed as anti-Mtb drug candidates [3]. Identification of protein-nucleoside ligand couples, and understanding the basis for functional molecular interactions, are complicated by obscurities in the relationships between gene sequences and ligand binding specificity. For example, enzymes sharing sequence similarity can exhibit differences in nucleotide specificity [4 – 6], and the specific ligand-binding properties of individual proteins can be sensitive to small changes in gene sequence [7 – 9]. Aiming to complement bioinformatics approaches to protein function [10, 11], we describe a biochemical screen based on dye-ligand chromatography [12] that can be used to obtain evidence for specific protein-ligand interactions and to assess selectivity of protein-ligand interactions in the human pathogen *Mycobacterium tuberculosis*.

Affinity binding technologies provide a means to identify physical interactions between macromolecules and their small-molecule partners [13 – 15]. The technique of dye-ligand chromatography has been used to purify a wide variety of proteins [12]. Cibacron Blue F3GA binds many nucleoside-dependent enzymes and, by selective elution with salt or ligands such as NADH or AMP, these proteins can often be purified [12]. Here we use the selective elution of proteins from F3GA chromatography resin as a high-throughput assay for protein-ligand interactions. The premise in our approach is that if a ligand can trigger release of a protein adsorbed on the dye-column, the ligand probably interacts in a specific fashion with that protein. This approach was first applied to crude cell extracts; the specificity of ligand-protein interactions were then further examined using purified recombinant proteins.

IDENTIFICATION OF PUTATIVE NUCLEOSIDE-BINDING PROTEINS: LIGAND-SPECIFIC ELUTION OF NATIVE PROTEINS BOUND TO DYE-RESIN

When we applied a crude extract of soluble proteins from *Mycobacterium tuberculosis* (Mtb) to a column of F3GA resin, approximately 40% of the total protein remained adsorbed after extensive washing. This Mtb protein-loaded F3GA resin was used in our first ligand-specific elution screen for nucleoside/nucleotide-protein interactions, outlined in Figure 1, in which 6 to 20 potential ligands were applied successively to a given protein-loaded column, and the proteins eluted by each ligand were identified by standard proteomic tools [16]. Relatively low nucleoside concentrations (1 mM) and small volumes of eluting buffer (0.5 column bed volume) were employed, with the aim of identifying specific ligand-protein interactions.

Furthermore, the strong interaction of many proteins with F3GA, and the high concentration of F3GA in the resin, also served to enhance the stringency of this ligand-protein interaction assay.

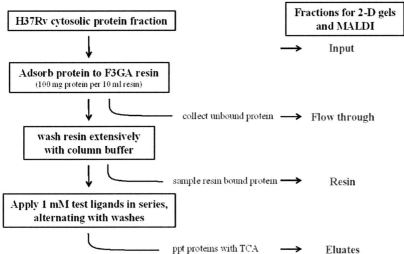

Figure 1. Purification of candidate nucleoside-ligand binding proteins from *M. tuberculosis* cell lysates. Affinity chromatography using immobilized Cibacron Blue F3GA was performed as outlined by Scopes [10], where Mtb proteins that bind F3GA are tested for ligand-specific elution by adding individual nucleosides or nucleotides to the column buffer. A crude cytosolic extract (100 mg) from Mtb strain H37Rv [http://www.cvmbs.colostate.edu/microbiology/tb/top.htm] was desalted over a Sephadex G-25 column and adsorbed to a 10 ml "Affigel" Blue Gel (Cibacron F3GA Blue) (BioRad) affinity column. The affinity column was washed extensively with column buffer (CB; 50 mM KH$_2$PO$_4$, pH 7.5, 1 mM MgCl$_2$ and 2 mM DTT) to remove unbound and low-affinity proteins prior to ligand elution. Approximately 40% of total cytosolic protein bound to the resin, as determined by Bradford assay. An aliquot of the resin-bound protein was extracted for subsequent 2D-gel analysis prior to elution (see Figure 2, top left panel); ~100 mg resin was extracted in 250 µl urea sample buffer (USB; 8% urea, 2% NP-40, 18 mM DTT), and the solubilized protein recovered in a spin column (Costar Spin-X, cellulose acetate membrane). Ligand-specific elution was carried out using 5 ml (one half column volume) of each ligand at 1 mM in CB. Ligands were applied in series, and the column was washed with 20 ml (2 column volumes) CB between ligands. Up to 20 different ligands were used to elute proteins from a single column. Columns were monitored using an in-line flow cell at 260 nm. Peak ligand fractions were pooled and the protein precipitated by addition of 100% iced TCA to a final concentration of 20%. Precipitated proteins were recovered by centrifugation, washed with acetone, and solubilized in 300 µl urea sample buffer. Recovered proteins were fractionated by 2-dimensional IEF-SDS-PAGE, using 13 cm pH 3 – 10 NL Immobilon gradient strips in the first dimension (Pharmacia Biotech IPGphor system, as per the manufacturer's instructions), and 15% SDS slab gels in the second dimension. Proteins were stained with Coomassie Brilliant Blue R250; excised protein spots were trypsinized in situ; recovered peptides were analyzed by MALDI-TOF MS, and peptide masses were matched to predicted proteins in the Mtb genome [16].

We observed ligand-specific release of distinct assortments of individual proteins from the resin by many different ligands such as NAD and ATP (Fig. 2). Furthermore, the release of individual proteins from the Mtb protein-loaded resin was sensitive to small differences in the structure of individual ligands, such as the presence of methyl and phosphate groups on adenosine (Fig. 3).

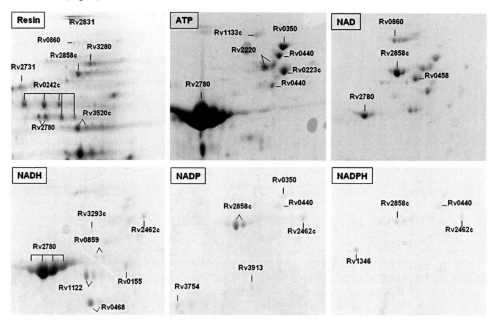

Figure 2. Ligand-specific elution of native Mtb proteins from Cibacron Blue F3GA resin. Two dimensional IEF-SDS-PAGE separation of Mtb cytosolic proteins that bind to the F3GA resin (top left panel), or are subsequently eluted by the indicated nucleotide at a concentration of 1 mM. Data are from three different experiments, as per figure 1. Only that portion of each gel corresponding to molecular mass ~100 to 35 kDa (top to bottom) and pI ~6 to 4.5 (left to right) is shown. Protein identities are indicated by locus tag number (see Table 1).

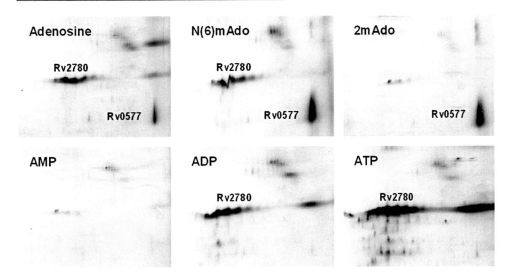

Figure 3. Selective elution of native Mtb proteins from F3GA resin by adenosine and adenosine analogs. Two dimensional IEF-SDS-PAGE separation of Mtb cytosolic proteins bound to F3GA and eluted successively by adenosine and related compounds. Ligands were applied to the column in the same order as the panels (from top left to bottom right, in order of increasing ligand mass), from a single experiment as per Figure 1. Rv0577 eluted with adenosine and both of the methyl-adenosine xenobiotics, but not with nucleotides. Conditions were as in Figure 2, except that proteins were stained with silver.

A summary of the predicted identities and functions (in Genbank) of the native Mtb proteins identified in different fractions described in Figure 1 are given in Tables 1 and 2. Table 1 lists 26 proteins that exhibited ligand-specific elution and in Table 3 these elution results are compared with the ligands associated with predicted protein functions available in public gene data bases. These tabulated results provide annotations of gene and predicted protein function at several levels. First, these data provide the first direct evidence that 28 Mtb proteins predicted from analysis of the Mtb genome exist *in vivo* (eight other proteins in Table 1 and 2, see footnotes). Second, the ligand-elution data provide specific annotations as to ligands that bind to these proteins. In the cases of Rv2671 and Rv3336c, for example, Table 3 shows that they bind NADPH and ATP, respectively. Table 3 contains many examples of consistency between nucleotide-elution data and the predicted ligand specificity based on gene sequence comparisons; Nineteen of the proteins in Table 3 were eluted by the predicted ligand or a very closely related one (e. g. NADP vs. NADH). Half (18/36) of the proteins identified in Tables 1 and 2 contain one or two Rossmann signatures that are associated with many nucleotide-binding proteins, including some with extensions beyond the $GX_{1-2}GXXG$ core [17] sequence.

Roberts, J.K.M. *et al.*

Table 1. Functional predictions for nucleotide-binding proteins from *M. tuberculosis* cytosolic extracts, identified by dye-ligand chromatography.

Locus tag	Protein name in GenBank	Rossmann motif
Amino acid transport and metabolism		
Rv1133c	methyltransferase metE	no
[a]Rv2220	glutamine synthetase glnA1	no
[a]Rv2780	secreted L-alanine dehydrogenase ald	*VI*GAGTAG*YNAA*
Rv3754	prephenate dehydrogenase tyrA	VLGLGLIGGSIM
Carbohydrate transport and metabolism		
Rv1023	phosphopyruvate hydratase eno	STGLGDEGGFAP
Rv1122	6-phosphogluconate dehydrogenase gnd2	MIGLGRMGANIV
Lipid transport and metabolism		
Rv0468	3-hydroxybutyryl-CoA dehydrogenase fadB2	*VV*GAGQMG*SGIA*
Rv0751c	3-hydroxyisobutyrate dehydrogenase mmsB	FLGLGNMGAPMS
Rv0824c	acyl-[acyl-carrier-protein] desaturase desA1	NMGMDGAWGQWVN
Rv0859	acetyl-CoA acetyltransferase fadA	no
Rv0860	fatty acid oxidation protein fadB	*VL*GAGMMG*AGIA*
		*IV*GYSGPAG*TGKA*
Nucleotide transport and metabolism		
Rv1843c	IMP dehydrogenase guaB1	AVGINGDVGAKAR
[a]Rv2445c	nucleotide diphosphate kinase ndkA	no
Other metabolism		
Rv0927c	short-chain dehydrogenase	no
Rv2671	hypothetical protein, possible ribD	no
Energy production and conversion		
Rv0155	NAD(P) transhydrogenase pntAa	VLGVGVAGLQAL
Rv0223c	aldehyde dehydrogenase	no
Rv0458	aldehyde dehydrogenase	QSGIGREGHQMM
Rv2858c	aldehyde dehydrogenase aldC	no
Rv3293	piperideine-6-carboxilic acid dehydrogenase pcd	no
[a]Rv3913	thioredoxin reductase trxB2	*VI*GSGPAG*YTAA*
Other cellular processes		
Rv1511	GDP-D-mannose 4,6 dehydratase gmdA	*IT*GITGQDG*SYLA*
Rv3336c	tryptophanyl-tRNA synthetase trpS	no
Poorly characterized		
Rv0484c	short-chain oxidoreductase	no
Rv0577	hypothetical protein	no
Rv1544	possible ketoacyl reductase	SAGFGTSGRFWE

Locus tag (Rv numbers) are gene identifiers for predicted open reading frames (ORFs) in the *M. tuberculosis* H37Rv genome (accessed through NCBI Entrez Gene at http://www.ncbi.nlm.nih.gov/ sites/entrez? db=gene). For each protein recovered by dye-ligand chromatography, the masses of tryptic peptides were determined by MALDI and matched to ORFs by ProteinProspector (http:// prospector.ucsf.edu/).
Rossmann motif shows which proteins contain the $GX_{1-2}GXXG$ core sequence characteristic of FAD, NAD(P)-binding proteins that form a Rossmann fold (BOLD), and those that also have the extended sequence shown to stabilize the fold structure (ITALIC; I or V in the 1st position shown, and A or G in the last) (Ref. 17). Note that Rv0860 contains 2 motifs.
[a] Indicates 4 proteins that have been previously studied (see also Table 2). The remainder are shown here, for the first time, to be expressed *in vivo*.

Table 2. *M. tuberculosis* proteins identified in this study for which no ligand-binding information was determined.

Locus tag	Protein name in GenBank	[b]Rossmann motif
Lipid transport and metabolism		
Rv0154c	probable acyl-CoA dehydrogenase fadE2	no
Rv0242c	3-ketoacyl-(acyl-carrier-protein) reductase fabG4	LVGNGSIGEGGR
		IAGIAGNRGQTNY
Rv2831	enoyl-CoA hydratase echA16	no
[a]Rv3280	propionyl-CoA carboxylase (beta chain 5) accD 5	VMGASGAVGFVYR
Energy production and conversion		
Rv3520	possible coenzyme F420-dependent oxidoreductase	ILGLGVSGPQVV
		LTGEGTTGLGKA
Molecular chaperones		
[a]Rv0350	molecular chaperone DnaK	no
[a]Rv0440	chaperonin groEL	VPGGGDMGGMDF
[a]Rv2031c	heat shock protein hspX	no
Rv2462c	trigger factor tig	no
Poorly characterized		
Rv2731	conserved hypothetical alanine/arginine rich protein	no

Proteins listed are those which bound to the affinity resin but did not elute with any of the test ligands (e.g., see Figure 1). In addition, all of the *Molecular Chaperones* were *ubiquitous*; present at low levels in all samples, including ligand-free washes.
[a, b] See Table 1 footnotes. Note that Rv0242c and Rv3520 each contain 2 *Rossmann motifs*.

Some of the ligand-protein combinations identified by ligand elution are unexpected or not predicted from sequence analyses and GenBank annotations (http://www.ncbi.nlm.nih.gov/Genbank/index.html). For example, Rv0577 has neither a Rossmann signature, nor any GenBank annotation suggesting nucleoside-dependent function. We therefore targeted such proteins for recombinant expression to examine ligand-protein interactions under conditions where the quantities and concentration of protein are well-defined, and the effects of different ligands on protein-dye interactions can be more readily compared. The method with crude extracts, just described, is a positive screen only; proteins that elute with ligands may not be picked up due to protein abundance or detection issues, and the order of eluting ligands can influence gel spot intensity and resolution for proteins which are substantially depleted by several related ligands. Moreover, some proteins retained on the column may be bound via other proteins in native complexes, rather than directly to the dye, while other proteins may have increased or unchanged affinity for the dye when bound to ligands.

Table 3. Predicted ligand-protein interactions for native *M. tuberculosis* proteins versus interactions observed by dye-ligand chromatography.

Locus tag	[a] Predicted interactions	[b] Ligand[s] causing elution from F3GA resin
Rv0155	NAD[P][H]	NADH
Rv0223c	NAD[P][H]	ATP
Rv0458	NAD[H]	N(6)mAdo, NAD
Rv0468	NAD[P][H]	NADH
Rv0484c	NAD[P][H]	NADPH
Rv0577	na	N(6)mAdo, 2mAdo >> Ado, FAD
Rv0751c	NAD[H]	NAD
Rv0824c	NADPH	NADPH
Rv0859	na	NAD[H]
Rv0860	NAD[H]	NAD[H]
Rv0927c	NAD[P][H]	NADP[H]
Rv1023	na	ATP
Rv1122	NADP[H]	NADH
Rv1133c	na	N(6)mAdo > ATP, GTP
Rv1511	NADP[H]	GTP
Rv1544	NAD[P][H]	NADPH
Rv1843c	NAD[H]	GMP
Rv2220	ATP/ADP/AMP	2mAdo >> N(6)mAdo > ATP, FAD
Rv2445c	NTP/NDP	ATP>GTP>>ubiquitous [c]
Rv2671	NADP[H]	NADPH
Rv2780	NAD[H]	NADH, ATP, cAMP >> ADP, Ado, N(6)mAdo
Rv2858c	NAD[H]	NADP > NADPH, NAD
Rv3293	NAD[P][H]	NADH
Rv3336c	ATP/AMP	ATP
Rv3754	NAD[H]	NADP
Rv3913	NADP[H], FAD	NADP

[a] *Predicted interactions* were compiled from biochemical studies of homologous proteins in other species, available in the NCBI, Prosite and BRENDA databases, and is not available (na) for all proteins.
[b] *Ligands causing elution* were observed experimentally for each protein. Relative spot intensities observed on 2-D gels are indicated by rank order, and do not reflect a systematic analysis of relative ligand affinity.
[c] See Table 2 footnote re *ubiquitous* proteins.

ANALYSIS OF THE SPECIFICITY OF LIGAND-PROTEIN INTERACTIONS WITH RECOMBINANT PROTEINS AND DYE-LIGAND CHROMATOGRAPHY

In the second ligand-specific elution screen for nucleoside/nucleotide-protein interactions, single recombinant Mtb proteins were adsorbed on multiple small (mg) aliquots of F3GA resin in spin columns, and assayed for elution by various nucleotides and nucleosides using one ligand per column, as outlined in Figure 4. The use of purified recombinant proteins permitted us to identify effects of a single variable (ligand structure) on the stability of the dye-protein complexes. We found clear differences in the release of a particular protein from the affinity resin by various ligands, shown for Rv0577 in Figure 5. We posit that greater amounts of eluted protein reflect stronger interaction between the protein and eluting ligand. Here, a stronger interaction between a protein and the eluting ligand could be due either to a direct interaction of the ligand at the dye-binding site of the protein, or binding at a site remote from the dye-binding site that causes conformational changes in the protein and weakens protein-dye interactions [12]. Further, only interactions that lead to a decrease in dye-binding affinity are expected to result in elution of a protein.

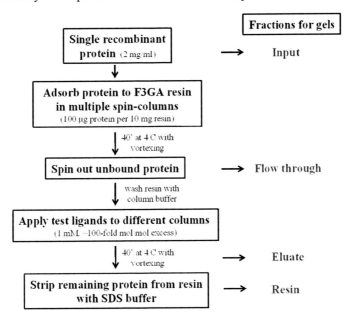

Figure 4. Analysis of nucleoside-ligand binding by recombinant proteins. Recombinant proteins were evaluated for their ligand-binding properties using a modified affinity elution chromatography protocol (22): individual proteins were diluted to 2 mg/ml in CB and adsorbed to multiple small aliquots of F3GA resin (100 μg protein per 10 mg resin] in 2 ml spin-columns (Costar Spin-X, cellulose acetate membrane). Binding was for 1 h at 4 °C with very gentle vortexing, followed by recovery of unbound protein (flow-through fraction) and washing of the resin (8 × 0.4 ml washes with CB); spin-columns were micro-centrifuged for 30 s at 13000 g, to recover fractions and change solutions. Individual spin-columns containing resin-bound proteins were then incubated (as for protein binding, above) with 50 μl 1 mM test ligand in CB, and the elution fractions recovered by centrifugation. Protein, which remained bound to the resin was recovered by heating at 95 °C for 5 min in 100 μl SDS sample buffer, and centrifugation (resin fraction). Aliquots of initial protein, spin-column flow-through and eluate fractions were diluted 1:1 with 2X SDS sample buffer. Equal proportions of all fractions (equivalent to 1 μg input protein) were loaded on 15% polyacrylamide one-dimensional mini-gels and stained with silver or Coomassie blue R250.

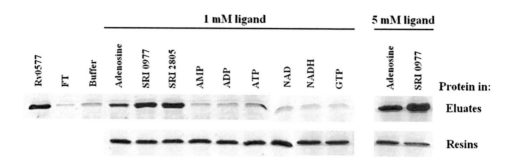

Figure 5. Ligand-specific elution of recombinant Mtb protein Rv0577 from Cibacron Blue F3GA resin. The protocol is described in Figure 4. Shown is the protein input (Rv0577), the protein that did not bind (FT), and the protein that was either released from F3GA resin by buffer or 9 different ligands (eluates) or remained bound to F3GA in the presence of the ligands (resins). SRI0977 = N(6)methyladenosine; SRI2085 = 2-methyladenosine. Silver staining shows release of Rv0577 from F3GA by nucleosides, but not by nucleotides.

The results with purified recombinant Mtb proteins such as Rv0577 (Fig. 5) are consistent with the observations from experiments with crude cytosolic extracts (Fig. 3; Table 3), which indicated that these proteins are nucleoside/nucleotide-binding proteins. Protein Rv0577 was eluted by the nucleoside adenosine as well as by methylated adenosine xenobiotics that are anti-Mtb drug candidates [3], whereas the common purine-containing mono- and di-nucleo-tides (AMP/ADP/ATP/NAD/NADH/GTP) did not promote release from the dye-resin (Fig. 5; Table 3).

CONCLUSIONS

We conclude that the two ligand-specific elution screens described here are useful for obtaining clues about the functions of members in the important class of nucleoside-binding proteins and their parent genes. Affinity methods are often effective in picking out functional classes of proteins from complex mixtures [e.g. 18], and our first dye-based screen followed this paradigm, sorting soluble proteins from a crude cell extract according to their ability to be released from the F3GA dye by millimolar concentrations of nucleosides (Figs. 2 and 3; Table 3). In addition to uncovering the natural ligands that nucleoside/nucleotide-binding proteins can associate with in cells, the dye-based screens can be extended to identify protein targets of nucleoside-analogs of interest as potential drugs [3]. Used in tandem, the two screens allow proteins to be selected from crude cell extracts as candidate nucleoside/ nucleotide-binding proteins and further examined, after cloning and expression, with respect to ligand preferences. Our results with proteins in crude extracts provided evidence that many proteins bind to multiple ligands (Table 3), findings that were confirmed in screens with pure, recombinant versions under conditions where differential release of protein from the dye resin can be assigned to differences in ligand structure (example shown in Figure 5).

The data from these combined screens are useful for confirming, refining or challenging gene annotations based on sequence analysis [e.g. 10, 11], and may also serve to complement the wealth of virtual ligand screening methodologies [19]. Furthermore, the ligand-binding data can guide protein crystallization efforts aimed at protein function via structure determination. In this context, information on ligands likely has considerable practical benefits, given that protein crystallization is often greatly improved by inclusion of ligands during crystal formation [20], and production of crystals suitable for analysis by X-ray diffraction is the principal bottleneck in structural genomics efforts today [21] (PSI target status website; http://sg.pdb.org/target_centers.html).

ACKNOWLEDGEMENTS

We thank Dr. W.W.P. Chang for help developing affinity sorting methods for cell extracts; Dr. J.T. Belisle for Mtb cytosol extracts (as part of NIH, NIAID Contract No. HHSN266200400091C to Colorado State University, Fort Collins, CO); Dr. R.C. Reynolds, Southern Research Institute, Birmingham, AL, for supplying the two methylated adenosine derivatives (SRI0977 and SRI2805); N. Maes and E. Z. Alipio Lyon for technical assistance. This work was in part supported by the LANL-UCR CARE program (STB-UC:06 – 29) and the NIGMS Protein Structure Initiative program (NIH U54 GM074946 – 01).

REFERENCES

[1]   Voet, D., and Voet, J.G. (2004) *Biochemistry*, 3rd Ed., Volume 1. J. Wiley & Sons, NJ.

[2]   Vilchèze C, Jacobs W.R. Jr. (2007) The mechanism of isoniazid killing: clarity through the scope of genetics. *Annu. Rev. Microbiol.* **61**:35 – 50.

[3]   Parker, W.B., and Long, M.C. (2007) Purine metabolism in *Mycobacterium tuberculosis* as a target for drug development. *Curr. Pharm. Des.* **13**:599 – 608.

[4]   Soderling, S.H., Bayuga, S.J., and Beavo, J.A. (1999) Isolation and characterization of a dual substrate phosphodiesterase gene family: PDE10A. *Proc. Natl. Acad. Sci. U.S.A.* **96**:7071 – 7076.

[5]   Joyce, M.A., Fraser, M.E., Brownie, E.R., James, M.N., Bridger, W.A., and Wolodko, W.T. (1999) Probing the nucleotide binding site of *Escherichia coli* succinyl CoA synthetase. *Biochemistry* **38**:7273 – 7283.

[6]   Tsybovsky, Y., Donato, H., Krupenko, N.I., Davies, C., and Krupenko, S.A. (2007) Crystal structures of the carboxyl terminal domain of rat 10-formyltetrahydrofolate dehydrogenase: implications for the catalytic mechanism of aldehyde dehydrogenases. *Biochemistry* **46**:2917 – 2929.

[7] Goward, C.R., and Nicholls, D.J. (1994) Malate dehydrogenase: a model for structure, evolution, and catalysis. *Protein Sci.* **3**:1883–1888.

[8] Eppink, M.H., Overkamp, K.M., Schreuder, H.A., and Van Berkel, W.J. (1999) Switch of coenzyme specificity of p-hydroxybenzoate hydroxylase. *J. Mol. Biol.* **292**:87–96.

[9] Wang, X., and Kemp, RG. (1999) Identification of residues of *Escherichia coli* phosphofructo-kinase that contribute to nucleotide binding and specificity. *Biochemistry* **38**:4313–4318.

[10] George, R.A., Spriggs, R.V., Bartlett, G.J., Gutteridge, A., MacArthur, M.W., Porter, C.T., Al-Lazikani, B., Thornton, J.M., and Swindells, M.B. (2005) Effective function annotation through catalytic residue conservation. *Proc. Natl. Acad. Sci. U.S.A.* **102**:12299–12304.

[11] Edgar, R.C., and Batzoglou, S. (2006) Multiple sequence alignment. *Curr. Opin. Struct. Biol.* **16**:368–373.

[12] Scopes, R.K. (1994) Protein *Purification: Principles and Practice*, 3rd Ed. Springer, NY.

[13] Makara, G.M., and Athanasopoulos, J. (2005) Improving success rates for lead generation using affinity binding technologies. *Curr. Opin. Biotechnol.* **16**:666–673.

[14] Erster, O., Eisenstein, M., and Liscovitch, M. (2007) Ligand interaction scan: a general method for engineering ligand-sensitive protein alleles. *Nat. Methods* **4**:393–395.

[15] Turnbull, J.E., and Field, R.A. (2007) Emerging glycomics technologies. *Nat. Chem. Biol.* **3**:74–77.

[16] Chang, W.W.P., Huang, L., Shen, M., Webster, C., Burlingame, A., Roberts, J.K.M. (2000) Patterns of Protein Synthesis and Tolerance of Anoxia in Root Tips of Maize Seedlings Acclimated to a Low Oxygen Environment, and Identification of Proteins by Mass Spectrometry. *Plant Physiol.* **122**:295–317.

[17] Kleiger, G., and Eisenberg, D. (2002) GXXXG and GXXXA motifs stabilize FAD and NAD(P)-binding Rossmann folds through Cα-H⋯O hydrogen bonds and van der waals interactions. *J. Mol. Biol.* **323**:69–76.

[18] Bieber, A.L., Tubbs, K.A., and Nelson, R.W. (2004) Metal ligand affinity pipettes and bioreactive alkaline phosphatase probes: tools for characterization of phosphorylated proteins and peptides. *Mol. Cell. Proteomics* **3**:266–72.

[19]   Villoutreix, B.O., Renault, N., Lagorce, D., Sperandio, O., Montes, M., and Miteva, M.A. (2007) Free resources to assist structure-based virtual ligand screening experiments. *Curr. Protein Pept. Sci.* **8**:381–411.

[20]   Vedadi, M., Niesen, F.H., Allali-Hassani, A., Fedorov, O.Y., Finerty, J.P.J., Wasney, G.A., Yeung, R., Arrowsmith, C., Ball, L.J., Berglund, H., Hui, R., Marsden, B.D., Nordlund, P., Sundstrom, M., Weigelt, J., and Edwards, A.M. (2006) Chemical screening methods to identify ligands that promote protein stability, protein crystallization, and structure determination. *Proc. Natl. Acad. Sci. U.S.A.* **103**:15835–15840.

[21]   Goh, C.-S., Lan, N., Douglas, S.M., Wu, B., Echols, N., Smith, A., Milburn, D., Montelione, G.T., Zhao, H., and Gerstein, M. (2004) Mining the structural genomics pipeline: Identification of protein properties that affect high-throughput experimental analysis. *J. Mol. Biol.* **336**:115–130.

[22]   Kim, C.-Y., Takahashi, K., Nguyen, T.B., Roberts, J.K.M., and Webster, C. (1999) Identification of a nucleic acid binding domain in eukaryotic initiation factor eIFiso4G from wheat. *J. Biol. Chem.* **274**:10603–10608.

Systems Chemistry, May 26th – 30th, 2008, Bozen, Italy

Systems Biology from Chemical Combinations

Joseph Lehár[*,1,2], Andrew Krueger[2], Grant Zimmermann[1], Alexis Borisy[1]

[1]CombinatoRx Incorporated,
245 First St, Cambridge, MA 02142, U.S.A.

[2]Boston University Bioinformatics/Bioengineering,
20 Cummington St, Boston, MA 02215, U.S.A.

E-Mail: *jlehar@combinatorx.com

Received: 28th October 2008 / Published: 16th March 2009

Abstract

Systematic testing of chemical combinations in cell-based disease models can yield novel information on how proteins interact in a biological system, and thus can make important contributions to biological models of those diseases. Such combination screens can also preferentially discover synergies with beneficial therapeutic selectivity, especially when used in high-order mixtures of more than two agents. These studies demonstrate the value obtainable from combination chemical genetics, and reinforce the growing realization that the most useful paradigm for a drug target is no longer a single molecule in a relevant pathway, but instead the set of targets that can cooperate to produce a therapeutic response with reduced side effects.

Introduction

Living organisms can be thought of as systems of interacting molecules, whose function and dysfunction would be better understood using a functional wiring diagram [1]. Such network models have already been used to predict the development of complex biological phenotypes [2], and will eventually help researchers gain a better understanding of multi-factorial diseases, identify novel therapeutic targets, and develop personalized treatments [3].

The staggering complexity of biological systems requires large and diverse sets of data on component connectivity and responses to system perturbations. The core topology of the network is determined using direct interactions between genes, proteins, and metabolites, and the models can then be refined using a biological system's responses to perturbations, such as drugs, mutations, or environmental changes. All of these approaches can associate network components with functional roles, even without direct interactions connecting them.

Combined perturbations shift the focus of genetic and chemical genetic studies from the functions of individual genes or proteins to interactions of those components within a biological system [4]. Large efforts to test the viability of double mutants are underway in bacteria [5], yeast [6], and nematodes [7]. Chemicals provide information that is distinct from and complementary to genetic perturbations, given the differences between how they modulate protein functions [8]. The advantages of chemical perturbations are that they can target a single domain of a multi-domain protein, allow precise temporal control that is critical for rapid-acting processes, can target orthologous or paralogous proteins enabling comparisons between species or redundant functions, and do not directly alter the concentrations of a targeted protein, thus avoiding indirect effects upon multi-protein complexes. Small molecules also lend themselves more readily to combination interventions, making them especially useful for integrating systems and chemical biology [9, 10]. Most combined perturbation studies with combinations have focused on exploring drug sensitivities across large sets of knockouts are also being undertaken in various model organisms [11, 12] and human cells [13]. However, these approaches are increasingly being extended to systematic testing of purely chemical combinations [10, 14]. Here we discuss our work that explores how combination studies can discover novel therapeutic treatments for complex disease systems, as well as to obtain biological information on the organization of functional components in those systems.

DISCOVERING COMBINATION DRUGS

One of the most powerful resources for discovering novel therapies is the set of existing drugs chemical probes with known biological activity. Over the past century, the pharmaceutical industry has had great success developing drugs with the following essential qualities: selective therapeutic activity with tolerable side effects; desirable absorption, distribution, metabolism, and excretion properties; and stable chemistry to facilitate storage, manufacturing, and distribution. Molecules with these properties are rare, and many disease relevant targets cannot be addressed by chemical inhibitors, so there are currently only ~3,000 approved drug ingredients and research probes which modulate 200 – 500 molecular targets within human biology [15]. Moreover, even the most specific small-molecule drugs can affect many different proteins, leading to off-target effects either through direct binding to secondary targets or through the downstream responses to these binding events.

Such secondary activities can occasionally be beneficial towards other diseases for which the drug was not originally developed, and indeed many successful drugs result from such repurposing efforts.

Phenotypic screening is a very effective method for discovering new opportunities provided by existing drugs [16]. Since the 1970s, the prevailing drug discovery paradigms have been target-based drug design and phenotypic screening. The target-based approach involves identifying a molecular target that is responsible for a disease phenotype and then system-atically testing a variety of chemical agents to see if they bind selectively to that target. By contrast, phenotypic screening tests chemical agents against a measurable quality of a whole organism's function, to identify those with a desirable response profile without necessarily having a detailed understanding of the drug's mechanism. While the target-based approach can discover primary target activities, secondary targets are almost exclusively found by phenotypic testing. Systematically testing many agents against phenotypes allows the biol-ogy to reveal which agents are likely to yield beneficial activities.

The opportunity for discovering useful off-target effects increases dramatically when drugs are used in combination [17]. Combinations are traditionally used to reduce toxicity, by sparing doses between two different drugs affecting the same disease function, to take advantage of complementary activities that independently improve the disease condition, and as the most effective response to drug-resistant pathogens, notably for viral and bacterial infections. Biological systems are robust, exhibiting a high degree of resilience that can compensate for a variety of attacks [18]. For example, mutation studies in many organisms [19, 20] find that only ~10 − 20% of genes affect viability when deleted from the genome. Combinations can overcome this robustness by targeting compensatory pathways [21], and many such cooperative relationships could work through known or yet undiscovered sec-ondary drug targets.

Chemical combinations need to be tested at varying drug doses, in order to find synergistic effects at unknown concentrations. This can be done with a "dose matrix", where a combi-nation is tested in all possible permutations of serially-diluted single agent doses (Fig. 1), or a fixed dose-ratio series, where component drugs are mixed at a high concentration and the mixture is tested in serial dilutions. Synergy is calculated by comparing a combination's response to those of its single agents [21], usually against the drug-with-itself dose-additive reference model [22]. Deviations from dose additivity can be assessed visually on an Iso-bologram (Fig. 1) or numerically with a Combination Index [23]. To capture synergies that can occur anywhere on full dose-matrix experiments, it is also useful to calculate a volume $V_{HSA} = \Sigma_{X,Y} \ln f_X \ln f_Y (I_{data} - I_{HSA})$ between the data and the highest-single-agent surface, normalized for single agent dilution factors f_X, f_Y, to quantify the strength of combination effects [21]. Both synergy scoring methods generalize readily to mixtures of three or more chemicals.

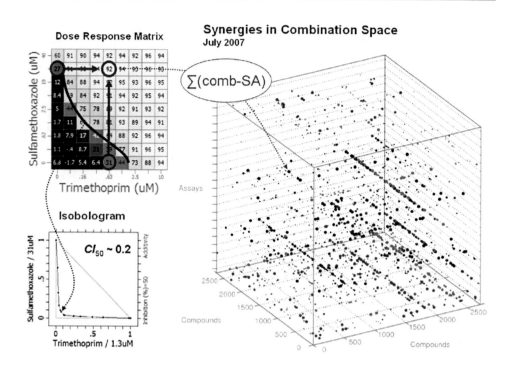

Figure 1. Dose matrix and synergy. Phenotypes (e.g., inhibition of *S. aureus* growth by Bactrim®) are measured for each drug, dosed in serial dilutions (bottom, left edges) and in all paired mixtures. Synergy is scored on differences between combined and single agent effects, while the isobologram shows the dose-sparing achieved over drug-with-itself dose-additivity. Using combination screening, we have discovered thousands of synergies in many disease areas. A cube showing 22 assays, each with colored symbols scaled to synergy, shows that despite some prolific drugs, strong synergies are rare.

CombinatoRx has been using a high throughput screening platform to explore combination therapies [24], by systematically testing combinations of ~3,000 agents in cell-based assays that preserve disease-relevant pathway complexity yet which are efficient enough to explore the vast space of combinations [21]. Using this platform, we have conducted dose-matrix screens in over ten disease areas to discover thousands of synergies (Fig. 1), many of which we have advanced into clinical trials for inflammation, diabetes and cancer indications. In addition, we have funded research programs utilizing our cHTS platform for drug discovery in neurodegenerative disease, infectious diseases, cystic fibrosis, muscular dystrophy, drug-device combinations and ophthalmology indications.

COMBINATIONS FOR SYSTEMS BIOLOGY

Dose matrix responses to combinations of chemical probes also can be used to determine mechanistic relationships between their targets [10]. Our therapeutic screens produce a wide variety of response surfaces, with distinct shapes for combinations having different known combination mechanisms. Many of these responses can be described by simple models that use the single agent curves to predict the combination effect (Fig. 2), and whose parameters provide quantitative measures of combination effects.

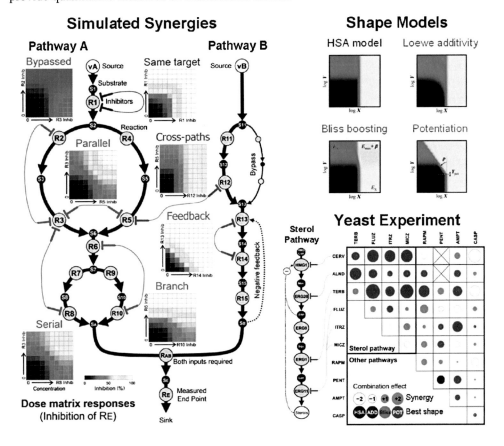

Figure 2. Synergy and connectivity. Simulated combination effects in a pathway of enzymatic reactions (grey enzymes, black substrates) with paired inhibitors (joined markers) depend on how the targets are connected. The effects can be described in terms of simple response shape models based on the single agent curves. A yeast experiment using sterol pathway inhibitors confirmed the expected response shapes in the pathway (symbol size scales with synergy, color shows the response model), and effects across pathways were weaker. Figure adapted from Lehár *et al.* 2007.

We established the relationship between synergy and connectivity by simulating a metabolic pathway as a series of linked Michaelis-Menten reactions with pairs of competitive inhibitors aimed at different targets with varying doses. We found that the shape and amount of synergy of each combination response depended on how the inhibitor pair's targets were connected in the pathway (Fig. 2). The predicted response shapes were robust to kinetic assumptions, parameter values, and nonlinear response functions, but were very sensitive to topological alterations in the target connectivity, such as branching, feedback regulation, or changing the type of junction at a branch point. The predicted shapes from these simulations were confirmed in a yeast proliferation experiment using drug combinations targeting sterol biosynthesis (Fig. 2), with further support from experiments on human cancer cells.

To explore the relationship between synergy and connectivity on a larger scale than can be addressed with dynamic pathway modeling, we are using flux-balance analysis (FBA) simulations of *E. coli* metabolism[25], with "minimization of metabolic adjustment" (MOMA) [26] to determine the impact of the perturbation upon bacterial growth. A variety of combination effects are observed, including many synergies in excess of the single agents, and a considerable variety of shapes that would have been indistinguishable with only knockouts. We estimated the total level of synergy by integrating the volume between the simulated data and the HSA model, $V_{HSA} = \Sigma_{X,Y} \ln f_X \ln f_Y (I_{data} - I_{HSA})$, summed over all positive concentrations X,Y and corrected for the dilution factors f_X and f_Y.

When the single agents are organized by their target location in the network, clear patterns emerge (Fig. 3). Most combinations produce no synergy, but there are strong synergies between adjacent enzymes within a pathway, and clear patterns of synergy between pathways, showing distinctions between the upper and lower ends of the pathway in some cases. These simulations demonstrate that genome-scale simulations of partially inhibited networks are tractable, and confirm that combination effects are strongly dependent on the connectivity of the perturbers' targets. Some of the response shapes clearly do not fit the shape models shown in Figure 2, so it is clear that we will need to broaden the reference set in order to capture the increased complexity of the system. Measuring connectivity between targets is considerably more challenging for a complex network like the *E. coli* FBA network,

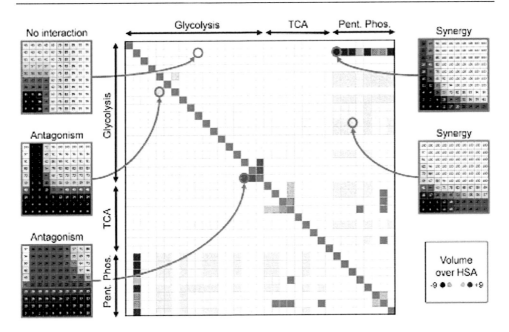

Figure 3. Simulated synergy. FBA/MOMA simulations of inhibited *E. coli* metabolism (central metabolism detail). The matrix shows synergy (V_{HSA}) for all pairs of targets sorted by pathway location. Most inhibitor pairs don't interact, but synergies and antagonisms show clear mechanistic patterns. The response surface shapes are more diverse than those shown in Figure 3.

We are also using flux balance analysis to model metabolism-dependent growth in *Saccaromyces cerevisiae* (yeast). This model organism has a fully sequenced genome, richly annotated functional pathways, and some of the most advanced network models both for metabolism and protein interactions [27, 28]. To provide an experimental basis for validation, we have screened all pairwise combinations of 60 chemical probes known to target metabolic enzymes in the FBA model, using an Alamar Blue metabolic readout as a proxy for cell growth (Fig. 4). Preliminary results from the combination screen are very encouraging. The observed responses are diverse, roughly ~4% of the tested combinations showed synergy (V_{HSA}) that were significantly different from the agent-with-self score distribution. The detected synergies are scattered throughout the combination space, but there are regions of consistent synergy between some pathways. In our preliminary studies, both individual combinations and synergy profiles confirm the results seen in our initial exploration of *Candida glabrata* [10]. We have performed FBA simulations of *S. cerevisiae* metabolism, which we will use for comparisons with our bacterial FBA models as well as with the yeast experimental results.

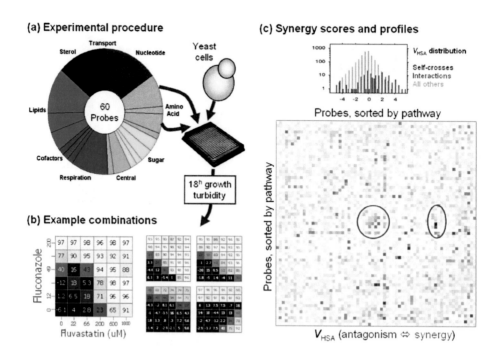

Figure 4. Yeast experiment. (**a**) Yeast cells were grown in pairwise combinations of 60 probes targeting metabolism. (**b**) Resulting responses confirm sterol synergies and produce diverse responses. (**c**) Analysis shows a 4% rate of interaction with synergies and antagonisms. Synergy profiles show both isolated synergies and coherent pathway interactions.

Higher order combinations (of three or more agents) can provide useful information about the overall complexity of a biological system. Randomly connected networks become increasingly sensitive to perturbations as the combination order is increased [29], and this trend is confirmed in more realistic network simulations using FBA approaches [30]. In principle, one would expect a trend towards more greater fragility to perturbation as the system is subjected to perturbations of increasing combination orders, due to their over-whelming any functional redundancy in the system (Fig. 5). From this it follows that by determining the "combination order of fragility" (COF), high-order perturbation experiments can be used to probe the robustness of a biological system [31]. We are experimenting with systematic high-order perturbation screens on bacterial proliferation assays, and pre-liminary results show that *Escherichia coli* survival networks have finite complexity that can be overcome with combinations at ~4th order (Fig. 5). This study also has identified a number of surprising synergies and antagonisms that arise at 3rd or 4th order without any suggestion at lower order. Such studies can be used to identify useful high-order synergies, to dissect a network in terms of its functional complexity, and to compare two systems (e. g., different bacterial species) in terms of their functional robustness.

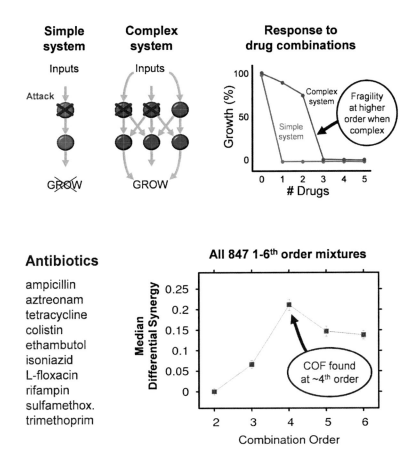

Figure 5. High order combinations. Single-pathway systems can be blocked by only one inhibitor, but redundant systems can function until all alternative routes have been blocked (upper). If perturbations are tested with increasing combination order, there will be incremental synergies until the functional redundancy has been reached, leading to a critical order above which the system is fragile to further attacks. A preliminary experiment testing all $1-6$th order combinations of ten diverse antibiotics (each with a 12-dose response curve) using an *E. coli* proliferation assay (lower) shows that the median differential synergy [31] across all combinations at each order peaks at ~4th order, suggesting a possible limit of functional redundancy.

Concluding Remarks

Our approach towards chemical combination studies is an efficient way to extract important information on the topology of biological networks. Individual combination responses show how targets are connected in ways that can't be resolved by genetic interactions, and analyzing synergy profiles across large combination experiments can be used to determine drug targets and pathway interactions. It even seems that systematic exploration of high-order combinations can provide information on the overall complexity of a system, in terms of its functional redundancy. Because our approach relies on phenotypic screening, the biases towards known pathways and mechanisms are minimized, increasing the opportunity for discovering totally novel biology in the process. In many ways, this provides an excellent empirical complement to systems biology efforts to generate network models to predict the behavior of complex diseases.

For drug discovery, combination approaches will help reinforce the growing realization that a useful paradigm for a therapeutic or bioengineering target is the set of nodes (e.g., metabolites, genes, proteins, or pathways) in a network that can selectively control the state of a biological system [32, 33]. In principle, the behaviour of biological systems should be controllable by individually adjusting the state of many components, and the precision with which the system can be manipulated should depend on the number of such state settings. Because of this increased precision, high-order perturbations are more likely than single agents to produce a therapeutic outcome without triggering toxic side-effects. In practice, therapeutic selectivity can result both from having more points of control and from the ability to reduce the doses of individual perturbers if they cooperate towards a beneficial endpoint. High order experiments can identify such selective synergies and determine the optimal number of ingredients. Although there will always be some conditions that are best treated by a single drug, high-order multi-target combinations represent a strategy that addresses the very complexity of biological systems.

References

[1]   Aderem, A. (2005) Systems biology: its practice and challenges. *Cell* **121**:511 – 3.

[2]   Davidson, E.H. *et al.* (2002) A genomic regulatory network for development. *Science* **295**:1669 – 78.

[3]   Weston, A.D. & Hood, L. (2004) Systems biology, proteomics, and the future of health care: toward predictive, preventative, and personalized medicine. *J. Proteome Res.* **3**:179 – 96.

[4]   Lehár, J., Stockwell, B.R., Giaever, G. & Nislow, C. (2008) Combination chemical genetics. *Nat. Chem. Biol.* **4**:674 – 81.

[5]   Butland, G., Babu, M., Díaz-Mejía, J.J., Bohdana, F., Phanse, S., Gold, B., Yang, W., Li, J., Gagarinova, A.G., Pogoutse, O., Mori, H., Wanner, B.L., Lo, H., Wasniewski, J., Christopolous, C., Ali, M., Venn, P., Safavi-Naini, A., Sourour, N., Caron, S., Choi, J.Y., Laigle, L., Nazarians-Armavil, A., Deshpande, A., Joe, S., Datsenko, K.A., Yamamoto, N., Andrews, B.J., Boone, C., Ding, H., Sheikh, B., Moreno-Hagelseib, G., Greenblatt, J.F., Emili, A. (2008) eSGA: *E. coli* synthetic genetic array analysis. *Nat. Methods* **5**:798–795.

[6]   Boone, C., Bussey, H. & Andrews, B.J. (2007) Exploring genetic interactions and networks with yeast. *Nat. Rev. Genet.* **8**:437–49.

[7]   Kamath, R.S. *et al.* (2003) Systematic functional analysis of the *Caenorhabditis elegans* genome using RNAi. *Nature* **421**:231–7.

[8]   Stockwell, B.R. (2004) Exploring biology with small organic molecules. *Nature* **432**:846–54.

[9]   Sharom, J.R., Bellows, D.S. & Tyers, M. (2004) From large networks to small molecules. *Curr. Opin. Chem. Biol.* **8**:81–90.

[10]   Lehár, J. *et al.* (2007) Chemical combination effects predict connectivity in biological systems. *Mol. Syst. Biol.* **3**:80.

[11]   Giaever, G. *et al.* (2004) Chemogenomic profiling: identifying the functional inter-actions of small molecules in yeast. *Proc. Natl. Acad. Sci. U.S A.* **101**:793–8.

[12]   Lum, P.Y. *et al.* (2004) Discovering modes of action for therapeutic compounds using a genome-wide screen of yeast heterozygotes. *Cell* **116**:121–37.

[13]   MacKeigan, J.P., Murphy, L.O. & Blenis, J. (2005) Sensitized RNAi screen of human kinases and phosphatases identifies new regulators of apoptosis and chemoresistance. *Nat. Cell Biol.* **7**:591–600.

[14]   Yeh, P., Tschumi, A.I. & Kishony, R. (2006) Functional classification of drugs by properties of their pairwise interactions. *Nat. Genet.* **38**:489–94.

[15]   Hopkins, A.L. & Groom, C.R. (2002) The druggable genome. *Nat. Rev. Drug Discov.* **1**:727–30.

[16]   Butcher, E.C. (2005) Can cell systems biology rescue drug discovery? *Nat. Rev. Drug Discov.* **4**:461–7.

[17]   Keith, C.T., Borisy, A.A. & Stockwell, B.R. (2005) Multicomponent therapeutics for networked systems. *Nat. Rev. Drug Discov.* **4**:71–8.

[18]   Hartman, J.L.t., Garvik, B. & Hartwell, L. (2001) Principles for the buffering of genetic variation. *Science* **291**:1001–4.

[19] Winzeler, E.A. *et al.* (1999) Functional characterization of the *S. cerevisiae* genome by gene deletion and parallel analysis. *Science* **285**:901 – 6.

[20] Fraser, A.G. *et al.* (2000) Functional genomic analysis of *C. elegans* chromosome I by systematic RNA interference. *Nature* **408**:325 – 30.

[21] Zimmermann, G.R., Lehár, J. & Keith, C.T. (2007) Multi-target therapeutics: when the whole is greater than the sum of the parts. *Drug Discov. Today* **12**:34 – 42.

[22] Greco, W.R., Bravo, G. & Parsons, J.C. (1995) The search for synergy: a critical review from a response surface perspective. *Pharmacol. Rev.* **47**:331 – 85.

[23] Chou, T.C. & Talalay, P. (1984) Quantitative analysis of dose-effect relationships: the combined effects of multiple drugs or enzyme inhibitors. *Adv. Enzyme Regul.* **22**:27 – 55.

[24] Borisy, A.A. *et al.* (2003) Systematic discovery of multicomponent therapeutics. *Proc. Natl. Acad. Sci. U.S.A.* **100**:7977 – 82.

[25] Reed, J.L., Vo, T.D., Schilling, C.H. & Palsson, B.O. (2003) An expanded genome-scale model of *Escherichia coli* K-12 (iJR904 GSM/GPR). *Genome Biol.* **4**:R54.

[26] Segrè, D., Vitkup, D. & Church, G.M. (2002) Analysis of optimality in natural and perturbed metabolic networks. *Proc. Natl. Acad. Sci. U.S.A.* **99**:15112 – 7.

[27] Ideker, T., Galitski, T. & Hood, L. (2001) A new approach to decoding life: systems biology. *Annu. Rev. Genomics Hum. Genet.* **2**:343 – 72.

[28] Zhang, L.V. *et al.* (2005) Motifs, themes and thematic maps of an integrated *Saccharomyces cerevisiae* interaction network. *J. Biol.* **4**:6.

[29] Agoston, V., Csermely, P. & Pongor, S. (2005) Multiple weak hits confuse complex systems: a transcriptional regulatory network as an example. *Phys. Rev. E. Stat. Nonlin. Soft Matter Phys.* **71**:051909.

[30] Deutscher, D., Meilijson, I., Kupiec, M. & Ruppin, E. (2006) Multiple knockout analysis of genetic robustness in the yeast metabolic network. *Nat. Genet.* **38**:993 – 8.

[31] Lehár, J., Krueger, A., Zimmermann, G. & Borisy, A. (2008) High-order combination effects and biological robustness. *Mol. Syst. Biol.* **4**:215.

[32] Kubinyi, H. (2003) Drug research: myths, hype and reality. *Nat. Rev. Drug Discov.* **2**:665 – 8.

[33] Kitano, H. (2007) A robustness-based approach to systems-oriented drug design. *Nat. Rev. Drug Discov.* **6**:202 – 10.

 Beilstein-Institut   Systems Chemistry, May 26th – 30th, 2008, Bozen, Italy

A Dynamical Supramolecular System for Medicinal Chemistry – A Step Towards Contiguous Chemical Spaces

Holger Wallmeier[1,*], Norbert Windhab[2], Gerhard Quinkert[3]

[1]Sossenheimer Weg 13, 65843 Sulzbach/Ts., Germany

[2]Evonik Röhm GmbH, Kirschenallee 41, 64293 Darmstadt, Germany

[3]Institut für Organische Chemie und Chemische Biologie,
Johann Wolfgang Goethe-Universität,
Max-von-Laue-Str. 7, 60438 Frankfurt, Germany

E-Mail: *holger.wallmeier@gmx.net

Received: 22nd February 2009 / Published: 16th March 2009

Abstract

A system based on pyranosyl-RNA (pRNA), a molecular scaffold which is able to self-assemble by Watson-Crick-like base pairing is presented. Molecular entities of very different types can be linked covalently to the components of the self-assembly scaffold to form conjugates which can be used in medicinal chemistry. Sets of conjugates with each of the different scaffold components define sublibraries for supramolecular assembly. Combining conjugates from different sublibraries, new (supra-)molecular entities with new properties can be formed by self-assembly in a systematic way. The supramolecular nature and the equilibrium of reversible self-assembly of the system ensure its dynamical behavior. In the presence of a molecular receptor, a complex system of equilibria exists, which allows controlling of the entire system. The dynamical properties of the system enable contiguous adaptation to changes of the conditions and offer new perspectives in obtaining structure/activity relationships.

INTRODUCTION

Oligonucleotides can self-associate specifically. Their interaction is based on hydrogen-bonding, base-stacking, and to some extent on entropy-related phenomena. The most famous example is the Watson-Crick-like pairing of DNA and RNA [1]. It results in supramolecular entities with characteristic properties which are of paramount importance for biological systems. The specificity of pairing, e.g., is crucial for transfer and handling of genetic information in living organisms. These very characteristics can be used to assemble supramolecular constructs made from suitable conjugates.

STRUCTURAL CONSIDERATIONS

DNA and RNA double strands have a characteristic helicity as a consequence of the backbone's topology (Figure 1). Due to the ribose unit in the backbone of RNA and DNA, the backbone has a certain twist which, together with the intra- and inter-strand interactions is responsible for the helicity.

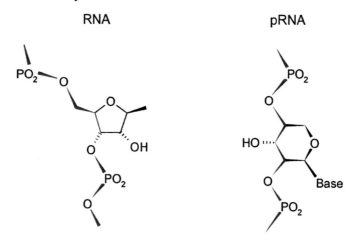

Figure 1. Schematic view of RNA and pRNA backbone (red) structure. Due to the 6-ring in the backbone, pRNA forms a more straight, alternating chain with valence angles close to the tetrahedron angle (109.47°) and shows less pronounced helicity, compared to RNA.

If one substitutes the ribose of the RNA backbone by pyranose, one obtains pyranosyl-RNA (pRNA) [2–4]. Whereas in RNA the backbone changes direction at the ribose units in a very characteristic way, all the atoms of the pRNA backbone are aligned in an alternating chain with valence angles close to the tetrahedron angle of 109.47°. As a consequence, the helicity of pRNA double helices is very shallow (see Figures 2 and 3); a pRNA double strand looks more like a warped ladder, rather than a helix.

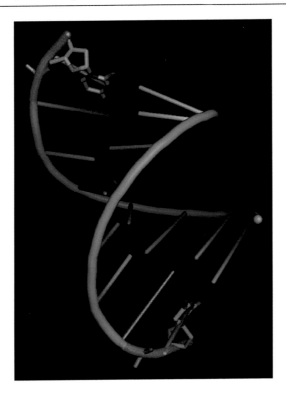

Figure 2. X-ray structure of the purine-pyrimidine alternating RNA double strand, r(GUAUAUA)d(C), with a 3'-terminal deoxy residue (PDB: 246D) [5].

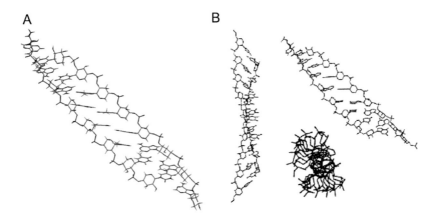

Figure 3. Structure of a pRNA double-strand. (**A**) Model structure of the pRNA double-strand p(AAAATTT)p(TTTTAAAA) based on MM/MD simulations with the AMBER 3.0 [4] force field. The simulation ensemble contained water and counter ions employing periodic boundary conditions at 298 K. (**B**) NMR structure of a pRNA double strand p(AAAATTT)p(TTTTAAAA) [7].

A set of three complementary strands, say **a**, **b**, and **c**, with **a** complementary to the first part of **c** and **b** to the second part, self-associate as depicted in Figure 4. If the sequences have been chosen accordingly, **c** can pair with **a** or **b** to form binary complexes **ac** and **cb**, which can then form the final ternary complex **acb** by pairing with the missing **b** or **a**. **c** is involved in all steps of pairing. The sequences have been chosen such that **a** and **b** cannot pair with each other, nor that any one of **a**, **b**, or **c** can pair with its own kind. In addition to Watson-Crick-like base-pairing by hydrogen-bonds, inter-strand π-π-interactions contribute to the stability of the supermolecules [8].

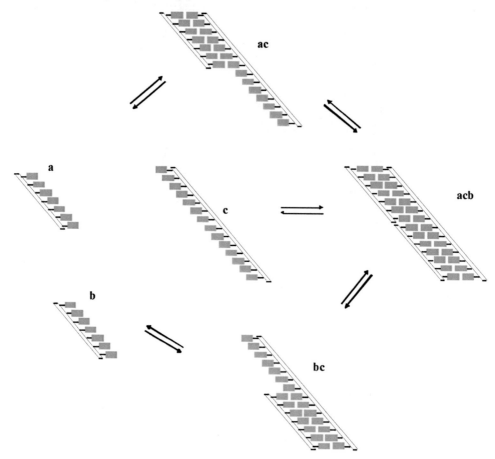

Figure 4. Self-association of complementary pRNA Strands.

Self-association dynamics

Dynamics of the self-association is controlled by the equilibria (1) through (5). A set of 5 independent equilibrium constants K controls the system. The subscripts i, j, and k are used to distinguish different sequences having the complementarity necessary.

$$c_i + a_j \leftrightarrow a_j : c_i \qquad K_{ji}^{ac} = \frac{[a_j : c_i]}{[a_j] \cdot [c_i]} \qquad (1)$$

$$c_i + b_k \leftrightarrow c_i : b_k \qquad K_{ik}^{cb} = \frac{[c_i : b_k]}{[c_i] \cdot [b_k]} \qquad (2)$$

$$a_j : c_i + b_k \leftrightarrow a_j : c_i : b_k \qquad K_{jik}^{acb} = \frac{[a_j : c_i : b_k]}{[a_j : c_i] \cdot [b_k]} \qquad (3)$$

$$c_i : b_k + a_j \leftrightarrow a_j : c_i : b_k \qquad K_{ikj}^{cba} = \frac{[a_j : c_i : b_k]}{[c_i : b_k] \cdot [a_j]} \qquad (4)$$

$$a_j + b_k + c_i \leftrightarrow a_j : c_i : b_k \qquad K_{jki}^{abc} = \frac{[a_j : c_i : b_k]}{[a_j] \cdot [b_k] \cdot [c_i]} \qquad (5)$$

It should be noted that (5) represents a synchronous association of **a**, **b**, and **c**. As a triple collision, however, it is much less likely than the other four associations.

The set of equilibria (1) through (5) can be represented graphically. Figure 5 shows the network of equilibria on the left hand side. The nodes represent the molecular species involved. Each connecting line represents the conversion of one molecular species into another one by association, or dissociation, respectively. The number of species involved is two along the colored lines, and four along the black lines. Given a certain initial amount of **a**, **b**, and **c**, the concentrations of free and paired molecular species vary accordingly.

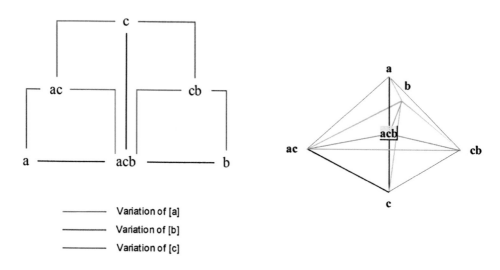

Figure 5. Network structure of the equilibria of self-association shown in Figure 4. A trigonal bipyramide is a simplex-like 3-dimensional representation of the equilibria. The corners represent single molecular species, the edges represent individual equilibria. The horizontal edge **ac-cb** represents an exchange-like process. It can be seen as a conflation of two distinct equilibria, **a** + **c** ↔ **ac**, and **b** + **c** ↔ **bc**, which are independent of each other.

In a three-dimensional representation, the nodes can be defined as the corners of a trigonal bipyramide with the connections forming the edges. It should be noted that the lengths of the edges depend on the relative amounts of **a**, **b**, and **c**. In other words, the symmetry of the bipyramide depends on the stoichiometric composition of the molecular system. As a consequence, whatever the positions of the equilibria are, the whole system is always found somewhere inside this type of simplex.

Stability of pRNA double strands

Figure 6 shows the melting curve of a 14 base pair pRNA double strand (see Figure 7) of a long strand and two short strands complementary to the first, respectively last 7 bases of the long strand. One can easily realize a well defined melting transition resulting from perfect pairing [8].

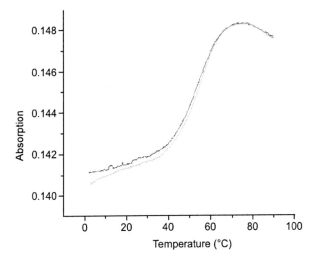

Figure 6. Melting curve of a three-component pRNA double strand [8]. The composition is given in Figure 7.

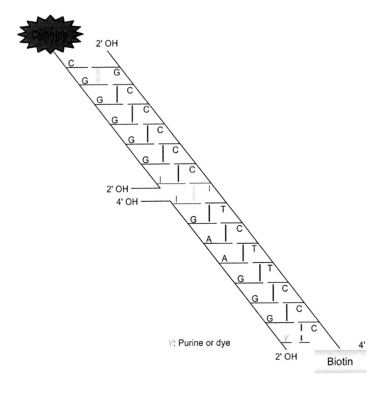

Figure 7. pRNA double strand composed of a 14-base strand and two complementary 7-base strands.

Conjugation

By the use of special linker bases, pRNA can be attached to other chemical entities forming conjugates [9, 10], denoted by **A**, **B**, and **C**. For all examples considered, conjugation does not interfere with base pairing (Figure 8), due to the uncritical topology of the pRNA double strand. Conjugates with peptides, carbohydrates, steroids, lipids, and even Fab fragments of antibodies have been prepared. Like the pure pRNA strands, also the conjugates self-associate according to the equilibria (1) through (5). Hence, on the basis of the three scaffolds, **a**, **b**, and **c**, three types of structural (sub)libraries can be created. By simply mixing components from each of the sublibraries, maximum possible diversity is obtained instantaneously.

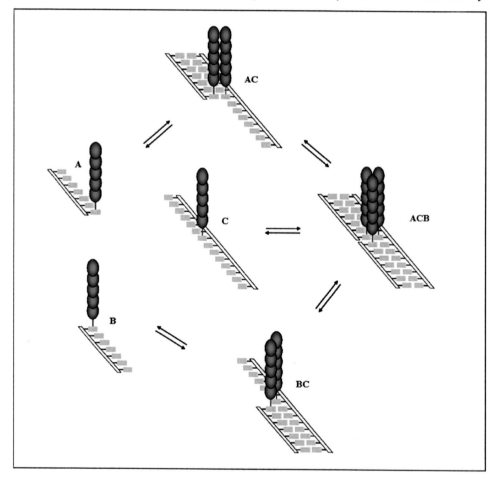

Figure 8. Pairing of pRNA strands is not affected by conjugation. Conjugation has been done with peptides, carbohydrates, steroids, lipids, and Fab fragments of antibodies.

Interaction with a receptor

The situation becomes more complex, if a receptor is added to the set of molecular species (Figure 9). Assuming that the receptor can bind to any of the molecular species, a number of additional equilibria appear (Figure 10). Bound to the receptor, paired pRNA conjugates can dissociate and associate, resulting in substitution processes that only exist in contact with the receptor. It is no longer guaranteed, however, that **A** and **B** can pair totally independent of each other with **C**, if they are in contact with a receptor.

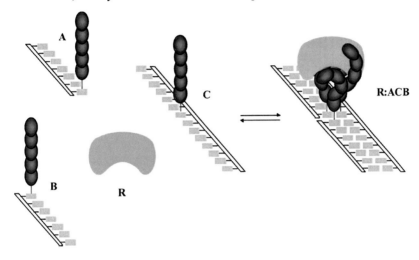

Figure 9. Adding a receptor to the self-associating system of pRNA conjugates.

In three dimensions the network of 8 molecular species and 28 equilibria can be represented by a cube. The connections of the 2-dimensional representation are now edges, face- and space-diagonals. Much in the sense of a simplex, the entire system always is found inside the cube, at a position, determined by the equilibrium constants. In the course of time, environmental influences on the system creates trajectories inside the cube, which can be taken as an indication of the system's intrinsic response flexibility.

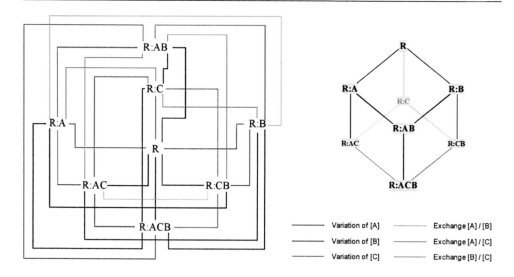

Figure 10. The equilibria of self-association in the presence of a receptor **R**. In contrast to the pure three-component system of pRNA conjugates, exchange reactions can occur in the presence of a receptor, since any of the molecular species can bind to the receptor. The 3-dimensional representation is given by a cube. The equilibria are now edges, face- and space-diagonals. The latter two kinds have been omitted for the sake of clarity. Again, the exact symmetry of this simplex depends on the stoichiometric ratios of the molecular species involved.

THE pRNA PAIRING SYSTEM AS A CARRIER OF ACTIVITY

Doubtlessly, conjugates can have biological activity. To study the behavior of the pRNA pairing system with respect to its effect on biological systems, three sublibraries with **A**-, **B**-, and **C**-type hexa-peptide conjugates have been synthesized. The pRNA sequences used were **a**: {CGGGGGN}, **b**: {NGAAGGG}, and **c**: {CCCTTCTNCCCCCC}. N is a Tryptamine nucleoside [9, 10] used as the site of conjugation. The peptides were chosen as discrete random sequences of the amino acids Cys, Glu, Phe, His, Lys, Leu, Asn, Arg, Ser, Thr, and Trp.

Receptor interaction analysis

The results of an interaction analysis with an enzyme and a particular triple Aρ, Bσ, and Cτ, based on surface plasmon resonance (Biacore instrument) have been combined with the results of a fluorescence-based inhibition experiment with the same species. ρ, σ, and τ denote different peptide sequences. For the interaction analysis a biotinylated conjugate of type **C** had been used to immobilize the molecular species **C**, **AC**, **BC**, and **ACB** on a chip, coated with streptavidin. For the measurement of activity a substrate of the enzyme labeled with a fluorescence dye was used that upon cleavage by the enzyme was activated and could be detected quantitatively by photometry. Figure 11 shows the correlation of the results. Starting from pure **C**, one sees that **B** lowers the affinity for the receptor **R**, but is important for activity. On the other hand, **A** has a significant effect on the affinity, but only a modest contribution to activity. Obviously there is a synergistic effect in the activity of the ternary complex **ACB**.

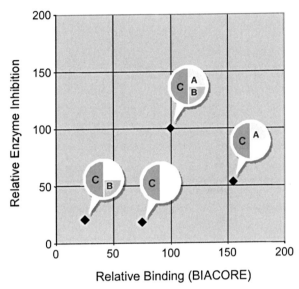

Figure 11. Correlation of affinity and activity data for a triple of conjugates **A**, **B**, and **C** from binding and inhibition experiments with an enzyme. Comparing **C** with **AC** and **CB** it can be seen directly that affinity and activity are conveyed by different components. In addition, a synergistic effect of the ternary complex **ACB** is evident. Data have been normalized relative to **ACB**.

The measurements have clearly shown that affinity and activity can be widely independent dimensions in pharmacological structure/activity relationships. Single-molecule entities are hardly able to resolve this correlation.

Enzyme inhibition assay

Figure 12 shows the result of an enzyme inhibition screening in the format of 16/96-well microtiter plates grouped in 4 columns and 4 rows. In $3 \, \mu M$ aqueous solution each well contained one of 8 different **A** components as indicated on the left margin, and one of the 11 different **B** components as indicated on the top margin. In addition, each well contained one of 1308 different **C** components. In the remaining wells, the individual **A** and **B** components, and the solvent with and without buffer were given as controls. In addition, the enzyme together with a fluorescence-labeled substrate was given into each well. In most of the wells the enzyme could cleave the substrate and fluorescence of the activated fluorophore was found. In the case of inhibition only little or no fluorescence was detected. In Figure 12 the color-coded hypersurface is shown that represents the degree of inhibition obtained as an average of two independent experiments. Surprisingly, not only single hits were found, but also extended areas of active combinations **Aρ**, **Bσ**, and **Cτ**. For instance, most combinations containing **A4** showed some activity. Obviously, however, this activity is lost in some of the plates due to the respective **C** components. It should be noted that neither any of the individual components, nor buffer and solvent used did show any notable inhibitory activity.

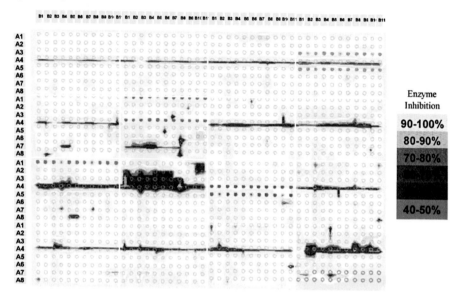

Figure 12. Enzyme inhibition assay with three sublibraries of pRNA peptide conjugates. Components **A** and **B** are indicated on the top and left margin. Component **C** is different in each well. The highest inhibitory activity ($IC_{50} = 23$ nM) was found in the combination at position **A8/B11** in the third row of the fourth column.

The strongest activity was found on the plate in column 4, row 3 at the position **A8/B11**. In a separate measurement an IC_{50} value of 23 nM was found.

In principle, the result in each well can result from four different scenarios. Given the substrate **S**, competitive inhibition (**ACB:R**), uncompetitive inhibition (**ACB:R:S**), mixed inhibition (**ACB:R + ACB:R:S**), and substrate capture by the conjugates can occur. The intra-conjugate interactions can be cooperative, as well as anti-cooperative. Furthermore, the guiding influence of component **C** for the self-assembling system may be weakened in contact with the receptor.

The second strongest inhibitor was found in the plate of the second column and the third row at position **A3**/**B1**. With this combination an enzyme inhibition experiment with variation of the concentrations of **A** and **B** was made. Figure 13 shows the hypersurface of residual enzyme activity as a function of the concentrations [**A**] and [**B**]. [**C**] was kept constant throughout the experiment. It is surprising that the stoichiometric 1:1:1 composition of the self-associating system is found just on top of a ridge, separating a flat area of the hypersurface from a descending slope. Hence, further increasing both, [**A**] and [**B**] leads to an increase in activity. One conclusion to be drawn is that probably not a single molecular species is responsible for the activity, which surely has some interesting consequences for pharmacokinetics, as well as pharmacodynamics.

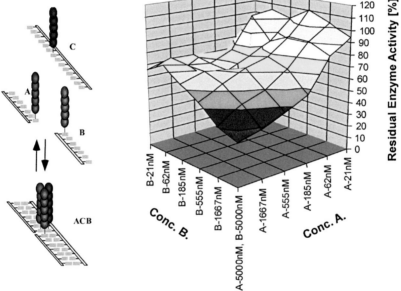

Figure 13. Variation of the stoichiometry of the self-associating system of pRNA peptide conjugates in an enzyme inhibition experiment. The [**A**] = [**B**] = [**C**] = 550 nM composition is on top of the yellow ridge in the center of the diagram. Increasing the concentrations of **A** and **B** leads to stronger inhibition of the enzyme.

Contiguous chemical structure spaces

Given the dynamical properties of the self-associating system of pRNA conjugates, one can imagine the unique possibility of gradual transitions from one structure class of conjugates to another one. By enriching the system **A**, **B**, and **C** with, *e. g.* a component **B'**, belonging to another structure class one can observe the change in the response of, a biological target system. The related equilibria guarantee that at all times the corresponding mixture of **B** and

B', **BC** and **B'C**, as well as **ACB** and **ACB'** is present. Some questions still remain to be answered. For instance, the existence of possible second-order phase transitions in the transition between two structure classes has to be elucidated. The possible impact of such phase transitions on the response of biological systems is yet another topic to be studied.

CONCLUSION

A self-associating system like the pRNA conjugates described above offers a very elegant way of forming supermolecules in a well-defined manner. It can be seen as a mounting base for a large variety of structural classes. Advantages for medicinal chemistry are emerging. However, there is no obvious reason that such dynamical supramolecular systems could not be used successfully in other areas like, *e. g.* material sciences, molecular electronics, or catalysis.

With kind permission of Wiley-VCH, Figures 4, 5, and 8 – 13 have been taken from [11].

REFERENCES

[1]   Watson, J.D., Crick, F.H.C (1953) *Nature* **171**:737.

[2]   Pitsch, S., Wendeborn, S., Jaun, B., Eschenmoser, A. (1993) Pyranosyl-RNA (p-RNA). *Helv. Chim. Acta* **76**:2161.

[3]   Schlönvogt, I., Pitsch, S., Lesneur, C., Eschenmoser, A., Jaun, B., Wolf, R.M. (1996) Pyranosyl-RNA (p-RNA): Duplex Formation by Self-pairing. *Helv. Chim. Acta* **79**:2316.

[4]   Bolli, M., Micura, R., Pitsch, S., Eschenmoser, A. (1997) Pyranosyl-RNA: Further Observations on Replication. *Helv. Chim. Acta* **80**:1901.

[5]   Wahl, M.C., Ban, C., Sekharudu, C., Ramakrishnan, B., Sundaralingam, M. (1996) Structure of the purine-pyrimidine alternating RNA double helix, r(GUAUAUA)d(C), with a 3'-terminal deoxy residue. *Acta Crystallogr.* D **52**:655.

[6]   Weiner, S.J., Kollman, P.A., Case, D.A., Singh, U.C., Ghio, C., Alagona, G., Profeta, S., Weiner, P. (1984) *J. Amer. Chem. Soc.* **106**:765.

[7]   Ilin, S., Schlönvogt, I., Ebert, M.-O., Jaun, B., Schwalbe, H. (2002) Comparison of the NMR Spectroscopy Solution Structures of Pyranosyl-RNA and its Nucleo-δ-peptide Analogue. *ChemBioChem* **3**:93.

[8]   Hamon, C., Brandstetter, T., Windhab, N. (1999) Pyranosyl-RNA Supramolecules Containing Non-Hydrogen Bonding Base-Pairs. *Synlett.* **S1**:940.

[9]   Windhab, N. (2001) Final Report of the BMBF Project No. 0311030, Projektträger Jülich.

[10]   Miculka, C., Windhab, N., Quinkert, G., Eschenmoser, A. Novel Substance Library and Supramolecular Complexes Produced Therewith. PCT Int. Appl. WO 97/43232. *Chem. Abstr.* 1998, **128**:34984.

[11]   Quinkert, G., Wallmeier, H., Windhab, N., Reichert, D.(2007) Chemistry and Biology – Historical and Philosophical Aspects. *In*: Chemical Biology. From Small Molecules to Systems Biology and Drug Design (Eds. Schreiber, S.L., Kapoor, T.M., Wess, G.), Wiley-VCH, Weinheim.

Systems Chemistry, May 26th – 30th, 2008, Bozen, Italy

"Promiscuous" Ligands and Targets Provide Opportunities for Drug Design

Gisbert Schneider[1,*] and Petra Schneider[2]

[1] Johann Wolfgang Goethe-University,
Siesmayerstr. 70, 60323 Frankfurt am Main, Germany,

[2] Schneider Consulting GbR,
George-C.-Marshall Ring 33, 61440 Oberursel, Germany

E-Mail: *gisbert.schneider@modlab.de

Received: 9th February 2009 / Published: 16th March 2009

Abstract

Flexibility, structural adaptability and non-selective pharmacophoric features of ligands and macromolecular receptors are major challenges in molecular design. These properties can lead to promiscuous binding behavior and often prevent modeling of reasonable structure-activity relationships. We have analyzed neighborhood behavior of bioactive ligands and protein binding pockets in terms of molecular shape and pharmacophoric features. It turned out that there exist certain relationships between the shape and the buriedness of a protein pocket and its ability to accommodate a small molecule ligand. The self-organizing map concept and clustering techniques is presented as a means for predicting potential bioactivities of ligands, and "de-orphanizing" of drugs and receptor proteins. Opportunities for "scaffold-hopping" and "re-purposing" are discussed in the context of systems chemistry.

Introduction

The notion of "frequent hitters" [1, 2], in particular bioactive compounds with a broad ("promiscuous") binding profile to multiple targets [3, 4], has led to systematic investigation of the requirements for ligand binding selectivity. Such approaches often include a distinction between primary drug targets and "anti-targets" [5, 6], preferred and undesirable substructural elements in lead structure optimization [7, 8] and pharmacophoric ligand features [9, 10]. More recently these methods have been complemented by network analyses of one-to-many and many-to-many relationships between ligands and targets on a systemic level that is accessible by chemical biology and chemogenomics [11 – 14].

Computational methods support lead structure design by predicting target/anti-target binding and pharmacological effects of screening compounds and lead candidates [15]. These methods are either receptor- (relying on a three-dimensional receptor model) or ligand-based (starting from known ligands), and ideally both concepts are applied in parallel [16]. Here, we address the prediction and utilization of ligand "promiscuity" from these perspectives. First, we briefly introduce the "self-organizing map" (SOM) as a computational tool for data visualization and compound encoding [17]. Then, we demonstrate its applicability to "scaffold-hopping" [18] and "re-purposing" [19, 20], as well as the "de-orphanizing" of drugs and target proteins [21, 22]. We conclude that understanding the reasons for binding promiscuity will offer rich opportunities for future drug design.

THE SELF-ORGANIZING MAP

Among the different visualization techniques that are available to the drug designer, the SOM or "Kohonen network" [23, 24] has found particularly successful applications in hit and lead discovery. Clustering and visualization of high-dimensional data distributions are the two main applications of the SOM. This is most relevant for drug design and target prediction, as compounds are usually represented in a high-dimensional pharmacophoric feature space. Such a data projection can be schematically visualized as a two-dimensional map as illustrated in Figure 1. In this representation, high-dimensional compound distributions can be analyzed by visual inspection of the two-dimensional SOM projection [25].

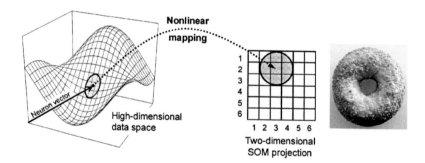

Figure 1. A SOM projection (*right*) can be used to visualize the data distribution in a high-dimensional data space (*left*) (figure adapted from ref. [22]). Here, a SOM containing $6 \times 6 = 36$ clusters (shown as a squared grid) is depicted. Local neighborhoods are conserved, that is, molecules that are close to each other on the SOM are also close in the original high-dimensional space (*e.g.*, a space spanned by many pharmacophoric features). The donut depicts a SOM forming a torus, *i.e.* an "endless two-dimensional space" [25].

The SOM belongs to a class of machine-learning systems which are optimized by "unsupervised learning" techniques [24]. According to this principle, so-called "neuron vectors" (cluster centroids) are positioned in the data space such that their distribution approximates the distribution of the data points. The term "unsupervised" means that no target information or activity values are required for the training process. By help of the SOM algorithm, a projection from a high-dimensional to a low-dimensional space can be obtained, so that clusters of isofunctional molecules can be identified by visual inspection (Figure 2). Briefly, unsupervised learning can be formulated as an iteration of a competitive (Step 2) and a cooperative (Step 3) step:

Step 0:   Define the SOM topology (e. g., rectangular toroidal) and number of clusters to be formed (*i. e.*, centroid vectors = neurons).

Step 1:   Choose a molecule from the high-dimensional descriptor space.

Step 2:   Determine the centroid vector ("winner neuron") that is closest to the molecule in high-dimensional space according to a distance metric.

Step 3:   Move the winner neuron and its neighboring neurons toward the data point.

Step 4:   Go to *Step 1* or terminate.

A software tool demonstrating this algorithmic concept is freely available from our public web server (http://gecco.org.chemie.uni-frankfurt.de/sommer/index.html).

SOM-BASED PREDICTION OF LIGAND BINDING PROFILES

In the attempt to predict the binding behavior of a small molecule, a first question to answer is how to represent the compound for mathematical analysis. Numerous descriptors have been suggested for this purpose [26]. Here, we used the topological pharmacophore descriptor CATS 2D, which encodes a molecule as a graph-based distribution of potential pharmacophoric points (hydrogen-bond donor or acceptor, lipophilic, positively or negatively charged atoms) and was shown to be suited for virtual compound screening and scaffold-hopping, *i. e.* the identification of isofunctional compounds with a different core structure [27, 28]. A literature-derived compilation of recently published drugs and lead compounds (COBRA version 8.6, 10840 compounds [29]) was encoded by the 150-dimensional CATS 2D descriptor and subjected to SOM clustering and projection (Figure 2). Here, the SOM was arranged as a 15 x 10 rectangular array with toroidal topology. Local clusters of compounds that bind to the same target are referred to as *activity islands* [29, 30]. Figure 2 shows examples of such activity islands. If compounds with a certain desired pharmacological activity, such as enzyme inhibition, or specific target binding, are found to be clustered on the SOM in an activity island, these clusters represent promising target areas for molecular design. Compounds molecules that fall in the desired target area on the SOM are candidates for synthesis and *in vitro* screening.

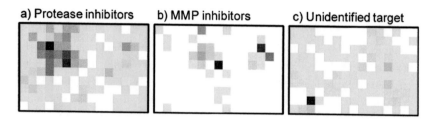

Figure 2. SOM projections of different ligand classes (annotated by target family). Color shading indicates ligand density (*white*: none, *yellow*: few, *magenta*: many). MMP: matrix metalloproteinases. Adapted from ref. [22].

Several hit and lead discovery projects have already successfully employed this method [30 – 35]. Notably, the location of ligands with unknown target protein ("orphan" ligands) on the SOM projection can even be used to identify target candidates by neighborhood analysis [20, 31]: From the known target-binding profiles of co-located drugs one can infer a similar activity for the orphan ligand. Such an example is given by compound **1** (Figure 3), which was predicted to bind to metabotropic glutamate receptor 1 (mGluR1) by selection of candidate compounds that co-located with known mGluR1 ligands on a SOM. This strategy led to the potent and subtype selective coumarine-derived mGluR1 antagonist **1** ($K_i = 24$ nM) [31], which was later developed into a potent lead series. It is important to keep note that SOMs were only used for first-pass compound filtering with the aim to design an activity-enriched focused library, and that subsequent chemical hit-to-lead optimization was still required. Pursuing a similar SOM-based concept, mGluR1 antagonists **2** and **3** were identified by picking structurally diverse compounds from a large screening compound database for bioactivity testing [34]. SOM-guided combinatorial library design yielded compound **4**, a selective antagonist ($K_i = 2.4$ nM) of purinergic receptor A_{2A} [35].

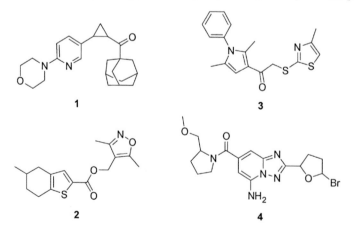

Figure 3. Examples of potent ligands that were identified using the SOM technique for virtual screening and target prediction.

These examples demonstrate the usefulness of ligand clustering by SOMs, with the aim to identify activity islands and exploit promiscuous binding behavior for the design of new bioactive compounds.

Ligand distributions are based on some sort of neighborhood relationship in chemical space, depending on the choice of the molecular descriptor (*e.g.*, pharmacophoric features, physicochemical properties, or substructure composition) [13, 36, 37]. As we have seen, it is possible to infer biological activity (*e.g.* receptor binding) from local ligand clusters. This concept is based on the analysis of a *single* or *few* compounds. An extension is to consider the distribution of all isofunctional ligands (defined, for example, as binding to the same drug target) in some chemical space, and compare *sets* of ligands with each other [38]. This enables the assessment of target similarity as measured by the similarity of the distributions of their respective ligands.

To illustrate the idea, we trained a SOM with the complete COBRA data (10,840 compounds represented by their CATS 2D descriptors). The SOM contained 150 neurons (clusters) arranged as a 15×10 rectangular array with toroidal topology. Then, this SOM was used to prepare individual projections for each target-specific ligand set, resulting in a total of 174 projections (*cf.* Figure 2 for example projections). The resulting patterns of the 174 individual ligand sets on the SOM were used as "target fingerprints" for the analysis of drug target similarity. In this study, the fingerprints consisted of $15 \times 10 = 150$ real-valued numbers (ligand density in each of the 150 SOM clusters). Similarity between the target fingerprints was expressed as Pearson correlation coefficient r, and graphically visualized as a target interaction network.

The network resulting from target relations with $r > 0.3$ is presented in Figure 4. We chose this threshold value because it represents a compromise between highly connected networks (obtained for lower values of r) and meagerly connected networks (larger values of r). 165 of the 174 targets analyzed show such a correlation to at least one other target *via* 334 edges. This suggests that most of the known drugs actually have multiple targets, which is in perfect accordance with earlier findings [39, 40]. It is noteworthy that several node clusters emerge, which can be attributed to similar ligand classes (not highlighted in Figure 4 for reasons of clarity). In addition, strongly connected nodes ("hubs") indicate targets by their respective ligands that have promiscuous binding features. Such an analysis complements receptor-based approaches for target comparisons and estimations of target "druggability" [41 – 45].

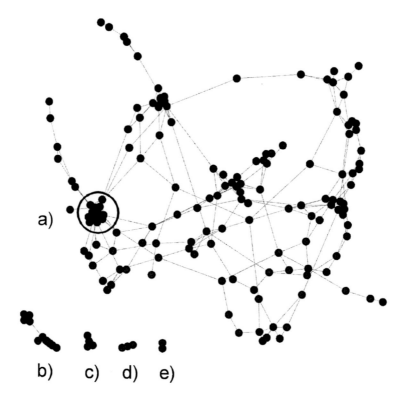

Figure 4. From ligand similarity to target similarity. Network representation of the relationship between ligand sets for 165 drug targets. The network structure results from correlation analysis of ligand properties (SOM-derived pharmacophoric features). Nodes represent drug targets, and the lines are scaled according to their pair-wise ligand similarity. Graphical representations of the target correlation network were prepared with the software Cytoscape 2.5.1 (http://www.cytoscape.org/). The network layout was optimized using the spring-embedded layout option provided within Cytoscape. Only targets with a pair-wise ligand correlation of $r > 0.3$ are shown. Selected targets, which potentially share similar ligand features:

a) *within the highlighted circle*: DAT (dopamine transporter DAT), NET (norepinephrine transporter), HTT (serotonine transporter), 5-HT3 (serotonine ion channel), dopamine receptor (GPCR), serotonine receptor (GPCR), histamine receptor (GPCR), potassium ion channel, nicotinic acetylcholine receptor (ion channel), monoamine oxidase (enzyme);

b) dihydrofolate reductase (enzyme), nitric oxide synthase (enzyme), RNA-polymerase (enzyme), P2Y (nuclotide receptor, GPCR), mGluR (metabotropic glutamate receptor, GPCR), AMPA (ionotropic glutamate receptor, ion channel), kainate receptor (ionotropic glutamate receptor, ion channel), P2X (nuclotide receptor, ion channel).

The targets contained in the small clusters **c)**, **d)**, and **e)** have unique ligand features that separate them from the majority of the known drug targets. These might therefore be particularly attractive for the development of novel selective lead compounds.

PREDICTION OF "DRUGGABLE" LIGAND BINDING POCKETS

The assessment of ligand promiscuity immediately poses the question whether there are preferable ligand-binding pockets that are "druggable" with the prospect of allowing for selective ligands to be found [41]. Several studies have been performed addressing this issue, and it turns out that comparably simplistic rules seem to allow for a distinction between druggable and non-druggable surface cavities [44, 45]. We have performed a comprehensive analysis of pocket shapes using known ligand-receptor complexes [45]. Figure 5 presents a SOM projection of liganded and unliganded pockets, which were described by their shape and buriedness [46]. Clearly, there are preferred shape-types that seem to govern the likeliness of ligand binding. This observation suggests that a limited set of preferred pocket shapes have evolved to accommodate substrates and effector molecules. Gaining a deeper understanding of these patterns will certainly help design not only ligands with a high "general likelihood" of binding by incorporation of certain substructure elements, but provide the necessary structural basis for the identification of target-selective ligands.

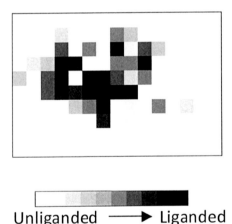

Unliganded   ⟶   Liganded

Figure 5. SOM giving the distribution of protein surface cavities that were encoded by a shape descriptor (adapted from ref. [45]). In total, 98 pockets with a ligand bound from structurally diverse proteins (distribution indicated by gray shading) were compared to 2,257 ligand-free cavities. A clear separation between liganded and unliganded pockets is visible.

An example of potential future applications of the pocket shape concept is presented in Figure 6. Angiotensin-converting enzyme (ACE) was analyzed by the software PocketPicker [46], resulting in many surface cavities (Figure 6, left panel). The largest pocket turned out to be the actual active site. Using *de novo* ligand design software, several inhibitor candidates were generated *in machina*, and docked into the active site pocket (Figure 6, right panel). The predicted binding modes are similar to Lisinopril, a known ACE inhibitor [47].

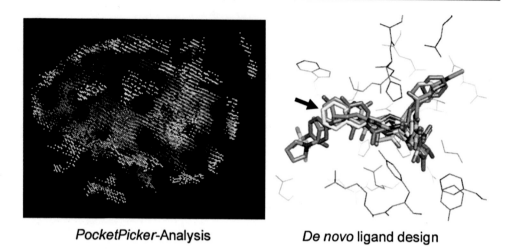

PocketPicker-Analysis   De novo ligand design

Figure 6. *Left*: Example of pocket analysis for an X-ray structure of ACE (Protein Data Bank identifier: 1o86 [48]). *Right*: Several *de novo* designed ligand candidates were docked into the active site. They exhibit essentially the same binding mode as the known ACE blocker Lisinopril (*gray-colored* co-crystal structure marked by the arrow).

This preliminary study demonstrates how we envisage systems chemistry might contribute to computer-assisted drug design in the future. Certainly, the SOM and other particular techniques presented here are not the only methods that can be applied. Much more advanced algorithms have been conceived. Computer science and in particular the various fields of technical engineering provide a rich source of pattern recognition and optimization strategies that may – when properly adapted to chemical applications – help understand and explore the potential of ligand "promiscuity" for rational drug design.

ACKNOWLEDGEMENTS

We are grateful to Martin Weisel, Ewgenij Proschak, Yusuf Tanrikulu, Björn Krüger, and Jan Kriegl for helpful discussion and technical assistance. This research was supported by the Beilstein-Institut zur Förderung der Chemischen Wissenschaften, Frankfurt.

REFERENCES

[1]   Roche, O., Schneider, P., Zuegge, J., Guba, W., Kansy, M., Alanine, A., Bleicher, K., Danel, F., Gutknecht, E.-M., Rogers-Evans, M., Neidhart, W., Stalder, H., Dillon, M., Sjögren, E., Fotouhi, N., Gillespie, P., Goodnow, R., Harris, W., Jones, P., Taniguchi, M., Tsujii, S., von der Saal, W., Zimmermann, G., Schneider, G. (2002) Development of a virtual screening method for identification of "frequent hitters" in compound libraries. *J. Med. Chem.* **45**:137 – 142.

[2]   Crisman, T.J., Parker, C.N., Jenkins, J.L., Scheiber, J., Thoma, M., Kang, Z.B., Kim, R., Bender, A., Nettles, J.H., Davies, J.W., Glick, M. (2007) Understanding false positives in reporter gene assays: *in silico* chemogenomics approaches to prioritize cell-based HTS data. *J. Chem. Inf. Model* **47**:1319 – 1327.

[3]   McGovern, S.L., Caselli, E., Grigorieff, N., Shoichet, B K. (2002) A common mechanism underlying promiscuous inhibitors from virtual and high-throughput screening. *J. Med. Chem.* **45**:1712 – 1722.

[4]   Morphy, R., Rankovic, Z. (2007) Fragments, network biology and designing multiple ligands. *Drug Discov. Today* **12**:156 – 160.

[5]   Klabunde, T., Evers, A (2005) GPCR antitarget modeling: pharmacophore models for biogenic amine binding GPCRs to avoid GPCR-mediated side effects. *ChemBioChem* **6**:876 – 889.

[6]   Raschi, E., Vasina, V., Poluzzi, E., De Ponti, F. (2008) The hERG K+ channel: target and antitarget strategies in drug development. *Pharmacol. Res.* **57**:181 – 195.

[7]   Hann, M.M., Oprea, T.I. (2004) Pursuing the leadlikeness concept in pharmaceutical research. *Curr. Opin. Chem. Biol.* **8**:255 – 263.

[8]   Hann, M.M, Leach, A.R, Harper, G. (2001) Molecular complexity and its impact on the probability of finding leads for drug discovery. *J. Chem. Inf. Comput. Sci.* **41**:856 – 864.

[9]   Sun, H. (2008) Pharmacophore-based virtual screening. *Curr. Med. Chem.* **15**:1018 – 1024.

[10]   Wolber, G., Seidel, T., Bendix, F., Langer, T. (2008) Molecule-pharmacophore superpositioning and pattern matching in computational drug design. *Drug Discov. Today* **13**:23 – 29.

[11]   Gregori-Puigjané, E., Mestres., J. (2008) A ligand-based approach to mining the chemogenomic space of drugs. *Comb. Chem. High Throughput Screen.* **11**:669 – 676.

[12]   Salemme, F.R. (2003) Chemical genomics as an emerging paradigm for postgenomic drug discovery. *Pharmacogenomics* **4**:257 – 267.

[13] Horvath, D., Jeandenans, C. (2003) Neighborhood behavior of *in silico* structural spaces with respect to in vitro activity spaces – a novel understanding of the molecular similarity principle in the context of multiple receptor binding profiles. *J. Chem. Inf. Comput. Sci.* **43**:680–690.

[14] Mestres, J., Martín-Couce, L., Gregori-Puigjané, E., Cases, M., Boyer, S. (2006) Ligand-based approach to *in silico* pharmacology: nuclear receptor profiling. *J. Chem. Inf. Model* **46**:2725–2736.

[15] Schneider, G., Baringhaus, K.-H. (2008) *Molecular Design – Concepts and Applications*. Wiley-VCH, Weinheim.

[16] Böhm, H.-J., Schneider, G. (Eds) (2000) *Virtual Screening for Bioactive Molecules*. Wiley-VCH, Weinheim.

[17] Bauknecht, H., Zell, A., Bayer, H., Levi, P., Wagener, M., Sadowski, J., Gasteiger, J. (1996) Locating biologically active compounds in medium-sized heterogeneous datasets by topological autocorrelation vectors: dopamine and benzodiazepine agonists. *J. Chem. Inf. Comput. Sci.* **36**:1205–1213.

[18] Schneider, G., Schneider, P., Renner, S. (2006) Scaffold-hopping: how far can you jump? *QSAR Comb. Sci.* **25**: 1162–1171.

[19] Chong, C.R., Sullivan Jr., D.J. (2007) New uses for old drugs. *Nature* **448**:645–646.

[20] Carley, D.W. (2005) Drug repurposing: identify, develop and commercialize new uses for existing or abandoned drugs. Part I. *Drugs* **8**:306–309.

[21] Cavasotto, C.N., Orry, A.J., Abagyan, R.A. (2003) Structure-based identification of binding sites, native ligands and potential inhibitors for G-protein coupled receptors. *Proteins* **51**:423–433.

[22] Schneider, P., Tanrikulu, Y., Schneider, G. (2009) Self-organizing maps in drug discovery: Library design, scaffold-hopping, repurposing. *Curr. Med. Chem.* **16**:258–266.

[23] Kohonen, T. (1982) Self-organized formation of topologically correct feature maps. *Biol. Cybern.* **43**:59–69.

[24] Kohonen, T (2001) *Self-Organizing Maps*. Springer-Verlag: Berlin.

[25] Zupan, J., Gasteiger, J. (1999) *Neural Networks in Chemistry and Drug Design*, Wiley-VCH: Weinheim.

[26] Todeschini, R., Consonni, V. (2000) *Handbook of Molecular Descriptors*, Wiley-VCH: Weinheim.

[27] Schneider, G., Neidhart, W., Giller, T., Schmid, G. (1999) "Scaffold-Hopping" by topological pharmacophore search: A contribution to virtual screening. *Angew. Chem. Int. Ed.* **38**:2894–2896.

[28] Fechner, U., Schneider, G. (2004) Optimization of a pharmacophore-based correlation vector descriptor. *QSAR Comb. Sci.* **23**:19–22.

[29] Schneider, P., Schneider, G. (2003) Collection of bioactive reference compounds for focused library design. *QSAR Comb. Sci.* **22**:713–718.

[30] Schneider, G., Schneider, P. (2004) Navigation in chemical space: Ligand-based design of focused compound libraries. In: *Chemogenomics in Drug Discovery* (Kubinyi, H., Müller, H., Eds), Wiley-VCH: Weinheim, pp. 341–376.

[31] Noeske, T., Sasse, B.C., Stark, H., Parsons, C.G., Weil, T., Schneider, G. (2006) Predicting compound selectivity by self-organizing maps: cross-activities of metabotropic glutamate receptor antagonists. *ChemMedChem* **1**:1066–1068.

[32] Yan, A. (2007) Application of self-organizing maps in compounds pattern recognition and combinatorial library design. *Comb. Chem. High Throughput. Screen* **9**:473–480.

[33] Hristozov, D., Oprea, T.I., Gasteiger, J. (2007) Ligand-based virtual screening by novelty detection with self-organizing maps. *J. Chem. Inf. Model* **47**:2044–2062.

[34] Renner, S., Hechenberger, M., Noeske, T., Böcker, A., Jatzke, C., Schmuker, M., Parsons, C. G., Weil, T., Schneider, G. (2007) Searching for drug scaffolds with 3D pharmacophores and neural network ensembles. *Angew. Chem. Int. Ed.* **46**:5336–5339.

[35] Schneider, G., Nettekoven, M. (2003) Ligand-based combinatorial design of selective purinergic receptor (A2A) antagonists using self-organizing maps. *J. Comb. Chem.* **5**:233–237.

[36] Bajorath, J. (2008) Computational analysis of ligand relationships within target families. *Curr. Opin. Chem. Biol.* **12**:352–358.

[37] Li, A., Horvath, S. (2007) Network neighborhood analysis with the multi-node topological overlap measure. *Bioinformatics* **23**:222–231.

[38] Schneider, G., Tanrikulu, Y., Schneider, P. (2009) Self-organizing molecular fingerprints: a ligand-based view on druglike chemical space and off-target prediction. *Future Med. Chem.*, in press.

[39] Keiser, M.J., Roth, B.L., Armbruster, B.N., Ernsberger, P., Irwin, J.J., Shoichet, B.K. (2007) Relating protein pharmacology by ligand chemistry. *Nat. Biotechnol.* **25**:197–206.

[40] Park, K., Kim, D. (2008) Binding similarity network of ligand. *Proteins* **71**:960–971 (2008).

[41] An, J., Totrov, M., Abagyan, R. (2004) Comprehensive identification of "druggable" protein ligand binding sites. *Genome Inform.* **15**:31–41.

[42] Zhang, Z., Grigorov, M.G. (2006) Similarity networks of protein binding sites. *Proteins* **62**:470–478 (2006).

[43] Kupas, K., Ultsch, A., Klebe, G. (2008) Large scale analysis of protein-binding cavities using self-organizing maps and wavelet-based surface patches to describe functional properties, selectivity discrimination, and putative cross-reactivity. *Proteins* **71**:1288–1306.

[44] Schalon, C., Surgand, J.S., Kellenberger, E., Rognan, D. (2008) A simple and fuzzy method to align and compare druggable ligand-binding sites. *Proteins* **7**:17775–1778.

[45] Weisel, M., Proschak, E., Kriegl, J.M., Schneider, G. (2009) Form follows function: Shape analysis of protein cavities for receptor-based drug design. *Proteomics* **9**:451–459.

[46] Weisel, M., Proschak, E., Schneider, G. (2007) PocketPicker: analysis of ligand binding-sites with shape descriptors. *Chem. Cent. J.* **1**:7.

[47] Krueger, B.A., Dietrich, A. Baringhaus, K.-H., Schneider, G. (2009) Scaffold-hopping potential of fragment-based de novo design: The chances and limits of variation. *QSAR Comb. Sci.*, in press.

[48] Natesh, R., Schwager, S.L., Sturrock, E.D., Acharya, K.R. (2003) Crystal structure of the human angiotensin-converting enzyme-lisinopril complex. *Nature* **421**:551–554.

BIOGRAPHIES

Timothy Clark

was born in southern England and studied chemistry at the University of Kent at Canterbury, where he was awarded a first class honors Bachelor of Science degree in 1969. He obtained his Ph.D. from the Queen's University Belfast in 1973 after working on the thermochemistry and solid phase properties of adamantane and diamantane derivatives. After two years as an Imperial Chemical Industries Postdoctoral Fellow in Belfast, he moved to Princeton University as a NATO Postdoctoral Fellow working for Paul Schleyer in 1975. He then followed Schleyer to the Institut für Organische Chemie of the Universität Erlangen-Nürnberg in 1976. He is currently technical Director of the Computer-Chemie-Centrum in Erlangen. His research areas include the development and application of quantum mechanical methods in inorganic, organic and biological chemistry, electron-transfer theory and the simulation of organic and inorganic reaction mechanisms. He is the author of 250 articles in scientific journals and two books and is the editor of the Journal of Molecular Modeling and founding director of Cepos InSilico Ltd., a company jointly owned by the Universities of Erlangen, Southampton, Portsmouth and Aberdeen.

Athel Cornish-Bowden

carried out his undergraduate studies at Oxford, obtaining his doctorate with Jeremy R. Knowles in 1967. After three post-doctoral years in the laboratory of Daniel E. Koshland, Jr., at the University of California, Berkeley, he spent 16 years as Lecturer, and later Senior Lecturer, in the Department of Biochemistry at the University of Birmingham. Since 1987 he has been Directeur de Recherche in three different laboratories of the CNRS at Marseilles. Although he started his career in a department of organic chemistry virtually all of his research has been in biochemistry, with particular reference to enzymes, including pepsin, mammalian hexokinases and enzymes involved in electron transfer in bacteria. He has written several books relating to enzyme kinetics, including *Analysis of Enzyme Kinetic Data* (Oxford University Press, 1995) and *Fundamentals of Enzyme Kinetics* (3 rd edition, Portland Press, 2004). Since moving to Marseilles he has been particularly interested in multi-enzyme systems, including the regulation of metabolic pathways. At present his main interest is in the definition of life and the capacity of living organisms for self-organization. In addition his principal areas of research, he has long had an interest in biochemical aspects of evolution, and his semi-popular book in this field, *The Pursuit of Perfection*, was published by Oxford University Press in 2004.

Antoine Danchin

trained as a mathematician and a physicist Antoine became an experimental microbiologist in the early seventies. The main goal of his research has always been to try and understand how bacterial genes can function collectively in the cell, as a pre-requisite to understand the path from commensalism to pathogenicity. To this aim, he started in 1985 a collaboration with computer scientists for evaluation of artificial intelligence techniques to the study of integrated problems in molecular genetics. This convinced him that it was time to investigate genomes as wholes, provided that an important effort in computer sciences was initiated in parallel. Early in 1987 he proposed that a sequencing program should be undertaken for *Bacillus subtilis*, the model of Gram positive bacteria. This proposal was actualized by an European joint effort on this genome, starting in 1988. The complete sequence has been published in 1997.

The first significant and unexpected discovery of this work was, in 1991, that many genes (at that time half of the genes) were of completely unknown function. As a further outcome of this work, it has been discovered that genomes are structures that are much more ordered than previously suspected, and that there probably exists a strong interaction between the organisation of the genes in the genome and the cell's architecture.

Antoine has published more than 500 articles and four books (300 in international scientific journals), including a book on the origin of life, and a book on genomes (*The Delphic Boat*, Harvard University Press, 2003). He has a continuous interest in philosophy, and in exchanges with other civilisations (formation of a Chinese-European University Without Walls in 1990), and has published many articles on subjects in epistemology and ethics. This was at the root of his interest to promote genome research in Hong Kong, where he has created the HKU-Pasteur Research Centre in 2000 at the Faculty of Medicine. He stayed in Hong Kong for three years and set up there a working seminar with the Department of Mathematics of Hong Kong University to discuss epistemological and ethical problems raised by the recent status of Biology in human knowledge.

Alexander Heckel

Born:	26th February 1972, Lindau (Lake Constance), Bavaria, Germany
1992 – 1997	Chemistry studies at the University of Constance, Diploma thesis with Prof. Dr. R. R. Schmidt (*Solid-Phase Synthesis of Oligosaccharides*)
1997 – 2001	Ph.D. thesis at the ETH Zurich with Prof. Dr. D. Seebach (*Enantioselective Heterogeneous Catalysis with TADDOL and Salen on Silica Gel*)
2001 – 2003	Postdoctoral Fellow at the California Institute of Technology with Prof. Dr. P. B. Dervan (Recognition of DNA with Minor Groove Binders)

2003 – 2007	Independent Studies at the University of Bonn Mentor: Prof. Dr. M. Famulok
since 2007	W2-Professor at the University of Frankfurt, Cluster of Excellence Macromolecular Complexes

Research Interests:
Photochemistry in living organisms, Nucleic acid nanoarchitectures

Scholarships and Awards:

1992 – 1996	Hundhammer-Fellowship (Bavarian Government)
1997	Prize for best diploma in chemistry at the University of Constance
1998 – 1999	Kekulé-Fellowship (Fonds der Chemischen Industrie)
2001 – 2003	Feodor Lynen-Fellowship (Alexander von Humboldt-Foundation)
2003 – 2006	Liebig-Fellowship (Fonds der Chemischen Industrie)
2006	Thieme Journal Award
since 2006	Emmy Noether Fellowship (DFG)

Martin G. Hicks

is a member of the board of management of the Beilstein-Institut. He received an honours degree in chemistry from Keele University in 1979. There, he also obtained his PhD in 1983 studying synthetic approaches to pyridotropones under the supervision of Gurnos Jones. He then went to the University of Wuppertal as a postdoctoral fellow, where he carried out research with Walter Thiel on semi-empirical quantum chemical methods. In 1985, Martin joined the computer department of the Beilstein-Institut where he worked on the Beilstein Database project. His subsequent activities involved the development of cheminformatics tools and products in the areas of substructure searching and reaction databases. Thereafter, he took on various roles for the Beilstein-Institut, including managing directorships of subsidiary companies and was head of the funding department 2000 – 7. He joined the board of management in 2002 and his current interests and responsibilities range from organizing the Beilstein Bozen Symposia with the aim of furthering interdisciplinary communication between chemistry and neighbouring scientific areas to Beilstein Open Access involving the publication of Open Access journals such as the *Beilstein Journal of Organic Chemistry*.

Douglas B. Kell

1996 – 1970	Top Scholar, Bradfield College, Berks
1975	B.A. (Hons) Biochemistry at St John's College, Oxford. (Class 2-[1] with Distinction in Chemical Pharmacology).
1978	Senior Scholar of St John's College, Oxford, M.A. (Oxon), D.Phil. (Oxon)
1978 – 1980	SRC Postdoctoral Research Fellow
1980 – 81	Postdoctoral Research Assistant
1981 – 1983	SERC Advanced Fellow
1983 – 1988	'New Blood' lecturer in Microbial Physiology, all at the Department of Botany & Microbiology, University College of Wales, Aberystwyth
1988 – 1992	Reader in Microbiology, Dept of Biological Sciences, UCW, Aberystwyth
	Founding Director, Aber Instruments Ltd, Science Park, Aberystwyth
1992	Personal Chair, The University of Wales
1997 – 2002	Director of Research, Institute of Biological Sciences, UWA
2001-	Founding Director, Aber Genomic Computing
2002-	EPSRC/RSC Research Professor of Bioanalytical Science, UMIST
2004-	The University of Manchester
2005-	Director, BBSRC Manchester Centre for Integrative Systems Biology (http://www.mcisb.org/)

Awards and Memberships

1986	Recipient of the Fleming Award of the Society for General Microbiology
1998	Aber Instruments received a Queen's Award for Export Achievement
2004	Royal Society of Chemistry Interdisciplinary Science Award
2005	FEBS-IUBMB Theodor Bücher prize. 2005 Royal Society/Wolfson Merit Award
2005	Royal Society of Chemistry Award in Chemical Biology
2006	Royal Society of Chemistry/Society of Analytical Chemistry Gold Medal
2000 – 2006	Member, BBSRC Council, BBSRC Strategy Board, NERC Environmental Genomics Committee
2007-	Member, BBSRC Bioscience for Industry Panel; Member, UKPMC Advisory Board
2007 – 2008	Member STFC Science Board

He has published over 350 scientific papers, 24 of which have been cited over 100 times (plus 3 over 90). His H-index is 52.

Carsten Kettner

studied biology at the University of Bonn and obtained his diploma at the University of Göttingen. in the group of Prof. Gradmann which had the pioneering and futuristic name – "Molecular Electrobiology". This group consisted of people carrying out research in electrophysiology and molecular biology in fruitful cooperation. In this mixed environment, he studied transport characteristics of the yeast plasma membrane using patch clamp techniques. In 1996 he joined the group of Dr. Adam Bertl at the University of Karlsruhe and successfully narrowed the gap between the biochemical and genetic properties, and the biophysical comprehension of the vacuolar proton-translocating ATP-hydrolase. He was awarded his Ph.D for this work in 1999. As a post-doctoral student he continued both the studies on the biophysical properties of the pump and investigated the kinetics and regulation of the dominant plasma membrane potassium channel (TOK1). In 2000 he moved to the Beilstein-Institut to represent the biological section of the funding department. Here, he is responsible for the organization of the Beilstein symposia, research (proposals) and publication of the proceedings of the symposia. Since 2004 he coordinates the work of the STRENDA commission and promotes along with the commissioners the proposed standards of reporting enzyme data. In 2007 he became involved in the development of a program for the establishment of Beilstein Endowed Chairs for Chemical Sciences and related sciences.

Joseph Lehár

Positions and Employment

1985 – 1987   Teaching Assistant, Massachusetts Institute of Technology, Cambridge, MA

1987 – 1991   Research Assistant, MIT, Cambridge, MA

1991 – 1994   Postdoc. Res. Asst., Institute of Astronomy, Cambridge University, Cambridge, UK

1994 – 2000   Research Associate, Harvard-Smithsonian Center for Astrophysics, Cambridge, MA

2000 – 2002   Research Scientist, Whitehead Institute, Center for Genome Research, Cambridge, MA

2002 – 2004   Team Leader, Computational Biology, CombinatoRx Inc, Cambridge, MA

2004-Present   Director, Computational Biology, CombinatoRx Inc, Cambridge, MA

Other Experience and Professional Memberships

1997–2000	Teaching Fellow, Harvard University, Cambridge, MA
1999	Visiting Scholar, Inst. of Astronomy, Cambridge University, Cambridge, UK
2000–2001	Consulting Scientist, NetGenics Inc., Cleveland, OH
2001–2002	Consulting Scientist, BioSift, Inc., Cambridge, MA
2002	Consulting Scientist, US Genomics, Inc., Woburn, MA
2002–Present	Adjunct Assistant Professor, Boston University, Boston, MA

Honors

1980	National Merit Scholarship Program, Finalist
1985	Brandeis University, Magna cum Laude with Highest Honors
1985	Brandeis University, Physics Faculty Prize
1990	Sigma Xi, MIT Chapter, Elected to membership
2000	Harvard University, CUE Certificate of Distinction in Teaching

Steven V. Ley

Steve Ley is currently the BP (1702) Professor of Chemistry at the University of Cambridge, and Fellow of Trinity College, Cambridge, UK. He studied for his Ph.D. at Loughborough University working with Harry Heaney and then carried out postdoctoral work in the USA with Leo Paquette at Ohio State University. In 1974 he returned to the UK to continue postdoctoral studies with Sir Derek Barton at Imperial College. He was appointed to the staff at Imperial College in 1975 and was promoted to Professor in 1983, and Head of Department in 1989. In 1990 he was elected to the Royal Society (London). In1992 he moved to his current post at Cambridge and from 2000–2002 he was also President of the Royal Society of Chemistry.

Professor Ley's work involves the discovery and development of new synthetic methods and their application to biologically active systems. The TPAP catalytic oxidant that is now used worldwide and cited extensively was one of his inventions. His group has published extensively on the use of iron carbonyl complexes, organoselenium chemistry, the use of microwaves in organic chemistry, biotransformations for the synthesis of natural products, and strategies for oligosaccharide assembly. To date more than 115 major natural products have been synthesised by the group. The group is currently developing new methods and techniques in particular the use solid-supported reagents in a designed sequential and multi-step fashion, and in combination with advances in the use of scavenging agents and catch and release techniques and for flow microreactor systems. Interesting advances have been made in the design of new catalysts especially for asymmetric synthesis.

Professor Ley has published over 630 papers and his achievements have been recognised by 32 major awards which include the Hickinbottom Research Fellowship, the Corday Morgan Medal and Prize, the Pfizer Academic Award, the Royal Society of Chemistry Synthesis Award for 1989, the Tilden Lectureship and Medal, the Pedler Medal and Prize, the Simonsen Lectureship and Medal and the Aldolf Windaus Medal of the German Chemical Society and Göttingen University, the Royal Society of Chemistry Natural Products Award, the Flintoff Medal, the Paul Janssen Prize for Creativity in Organic Synthesis, the Rhône-Poulenc Lectureship and Medal of the Royal Society of Chemistry and the Glaxo-Wellcome Award for Outstanding Achievement in Organic Chemistry.

Other notable accolades have been the Royal Society of Chemistry Haworth Memorial Lectureship, Medal and Prize and The Royal Society Davy Medal and the German Chemical Society August-Wilhelm-von Hofmann Medal together with the Pfizer Award for Innovative Science. In 2003 he was awarded the American Chemical Society Ernest Guenther Award in the Chemistry of Natual Products together with the Royal Society of Chemistry Industrially-sponsored award in Carbohydrate Chemistry, the Chemical Industries Association Innovation of the Year Award and the iAc Award; both jointly with AstraZeneca, Avecia and Syngenta. In 2004 he received the Messel Medal Lecture of the Society of Chemical Industry and the Alexander-von-Humboldt Award.

Recently his work has been acknowledged by the presigous Yamada-Koga Prize in 2005 and the Nagoya Gold Medal (Banyu Life Science Foundation International, Japan) and the Robert Robinson Award and Medal (Royal Society of Chemistry) in 2006 and in 2007 the Award for Creative Work in Synthetic Organic Chemistry (American Chemical Society) and the Karrer Gold Medal, University of Zurich.

Professor Ley has been the recipient of 74 named lectureships and given 303 special invited lectures. He has served on 42 national and international committees, 43 editorships and many industrial science advisory boards. He has been awarded honorary degrees and fellowships from the universities of Loughborough, Salamanca, Imperial College, London, Huddersfield and Cardiff.

Eric Meggers

Professional Preparation

1999 – 2002   The Scripps Research Institute
Postdoctoral fellow with Prof. Peter G. Schultz
a) Development of the concept of metallo-base pairing in DNA.
b) Phage display methodology for the ribosomal incorporation of unnatural amino acids into proteins.

1999   Univ. of Basel, Switzerland, Ph.D. in Organic Chemistry

1995	University of Bonn, Germany, Diploma in Chemistry

Appointments

2002 – June 2007   Assistant Professor, Department of Chemistry, University of Pennsylvania

July 2007 – present Professor, Philipps-University of Marburg

July 2007 – present Adjunct Assistant Professor, Wistar Institute, Philadelphia, USA

Awards and Honors

1994 – 1995	Fellowship from the Theodor-Laymann Foundation
1996	Award from the Heinrich-Hörlein-Memory Foundation
1995 – 1999	Fellowship from the Swiss National Science Foundation
1999 – 2000	Feodor Lynen Fellowship of the Alexander von Humboldt Foundation
2000 – 2001	Emmy Noether Fellowship from the DFG
2002	Camille and Henry Dreyfus New Faculty Award
2003	Synthesis-Synlett Journal Award
2006 – 2008	Alfred P. Sloan Research Fellow
2006	Camille Dreyfus Teacher-Scholar Award
2007	NanXiang Fellowship from Xiamen University, China

Justin K. M. Roberts

studied Agricultural and Forest Sciences at Oxford University (B.A., 1978) while performing research in plant biochemistry at ARC Letcombe Laboratory with D.T. Clarkson. On completion of his undergraduate training, he joined P.M. Ray's laboratory in Biological Sciences at Stanford University, primarily conducting research in O. Jardetzky's Stanford Magnetic Resonance Laboratory to measure intracellular pH gradients in plant cells (Ph.D. 1982). His post-doctoral work applying NMR to plant metabolism continued at Stanford until 1985, when he joined the Department of Biochemistry, University of California, Riverside as an assistant professor. He has been a Professor of Biochemistry at UCR since 1995, serving as Chairman 2001 – 4.

A predominant focus of his research has been nucleotide metabolism, from the pool sizes and physical states of nucleotides in plant cells, to rates of nucleotide turnover to, most recently, aspects of protein-nucleotide recognition in the context of functional and structural genomics in *Mycobacterium tuberculosis*.

Gisbert Schneider

studied biochemistry at the Free University (FU) of Berlin, Germany. From 1991 to 1994 he prepared his doctoral thesis on machine learning systems for peptide *de novo* design as a fellow of the Fonds der Chemischen Industrie. From 1994 to 1997 he performed post-doctoral research at the FU Berlin, the University of Stockholm, Sweden, the Massachusetts Institute of Technology, Cambridge, MA, USA, and the Max-Planck-Institute of Biophysics in Frankfurt, Germany. In 1997 he joined the pharmaceuticals division of F. Hoffmann-La Roche AG in Basel, Switzerland, where he became Head of Cheminformatics. Since 2002 he is a full professor of Chem- and Bioinformatics (Beilstein Endowed Chair for Cheminformatics) at Johann Wolfgang Goethe-University in Frankfurt, Germany, where he concentrates on the development and application of software methods for virtual screening and molecular design. He has published more than 150 scientific papers, books and patents and is author of two textbooks on molecular design. He is editor of the journal "QSAR and Combinatorial Science", and member of the editorial advisory boards of several journals in the field of chemical biology and drug discovery.

Peter H. Seeberger

received his Vordiplom in 1989 from the Universität Erlangen-Nürnberg, where he studied chemistry as a Bavarian government fellow. In 1990 he moved as a Fulbright scholar to the University of Colorado where he earned his Ph.D. in biochemistry under the guidance of Marvin H. Caruthers in 1995. After a postdoctoral fellowship with Samuel J. Danishefsky at the Sloan-Kettering Institute for Cancer Research he became Assistant Professor at the Massachusetts Institute of Technology in 1998 and was promoted to Firmenich Associate Professor of Chemistry with tenure in 2002. In June 2003 he assumed a position as Professor for Organic Chemistry at the Swiss Federal Institute of Technology (ETH) in Zurich, Switzerland and a position as Affiliate Professor at the Burnham Institute in La Jolla, CA.

Professor Seebergers research has been documented in over 170 articles in peer-reviewed journals, fourteen issued patents and patent applications, more than 90 published abstracts and more than 360 invited lectures. Among other awards he received the Technology Review Top 100 Young Innovator Award (1999), MITs Edgerton Award (2002), an Arthur C. Cope Young Scholar Award and the Horace B. Isbell Award from the American Chemical Society (2003), Otto-Klung Weberbank Prize.

In 2007 he was selected among "The 100 Most Important Swiss 2007" by the magazine "Schweizer Illustrierte", received the Havinga Medal, the Yoshimasa Hirata Gold Medal and the Körber Prize for European Sciences. Peter H. Seeberger is the Editor of the Journal of Carbohydrate Chemistry and serves on the editorial advisory boards of eleven other journals. He is a founding member of the board of the Tesfa-Ilg Hope for Africa Foundation that aims at improving health care in Ethiopia in particular by providing access to malaria vaccines and HIV treatments. He is a consultant and serves on the scientific advisory board of several companies.

The research in professor Seebergers laboratory has resulted in two spin-off companies: Ancora Pharmaceuticals (founded in 2002, Medford, USA) is currently developing a promising malaria vaccine candidate in late preclinical trials as well as several other therapeutics based on carbohydrates. i2chem (founded in 2005, Cambridge, USA) develops integrated microchemical systems based on silicon microreactors.

Holger Wallmeier

obtained his PhD 1982 in Theoretical Chemistry from the University of Bochum in the Group of Prof. Kutzelnigg with a thesis on relativistic quantum chemistry. In 1984 he took over a position at Hoechst AG in Frankfurt in the Scientific Computing Department of Corporate Research where he worked in computer-assisted drug design, simulation of bio-molecular systems, and software development. 1997 he founded the bioinformatics group at Hoechst Research & Technologies GmbH and was appointed head of the Core Technology Area Biomathematics in 1998.

Since than he has initialized and supervised numerous projects in bioinformatics, proteomics, expression analysis, development of bio-analytical software, and text mining. He has published a number of papers in quantum chemistry, structure/activity relationships, and protein structure prediction. 2003 he changed to Sanofi-Aventis doing consulting in the area of biocomputing for Sanofi-Aventis biotech investments and spin-outs of former Aventis Research & Technologies. Since 2008 he works as an independent consultant in the areas of scientific computing and bioinformatics.

Dave A. Winkler

studied chemistry, chemical engineering, and physics at the Monash and RMIT Universities in Melbourne. He completed a PhD in microwave spectroscopy and radioastronomy at Monash University in 1980 as a General Motors postgraduate fellow. He then worked as a tutor and senior research fellow, helping establish the computer-aided drug design group at the Victorian College of Pharmacy under Prof. Peter Andrews, in the early 80s. He subsequently worked as a senior research scientist with the Defence Science and Technology organization in Adelaide for two years, modelling energetic polymer properties.

He joined CSIRO Applied Organic Chemistry in Melbourne as a senior research scientist in 1985 where he worked on the design of biologically active small molecules as drugs and agrochemicals. He was awarded Australian Academy of Science traveling fellowships to Toshio Fujita's lab in Kyoto in 1988, and to Graham Richards' research group in Oxford in 1997. He has remained with CSIRO and his worked a common theme of the application of computational methods to understanding and designing novel small molecules with biological activity. Major projects during this time involved industry collaborations: development of HBV drugs with AMRAD, crop protection agents with Du Pont, and veterinary drugs with Schering Plough.

His research contributed to the formation of the start up companies such as Betabiotics. He was a member of the Biomolecular Research Institute, headed by Prof. Peter Colman who led the structural biology behind the flu drug Relenza. During has time, at CSIRO, a common theme of his work has been the development and application of Bayesian methods, particularly the Bayesian regularized artificial neural network, to the modelling of biological activity. During the last decade he has worked with a diverse complex systems science group in Australia and internationally. This aimed to understand how complexity concepts such as emergence, criticality, self-organization and self-assembly can contribute to the modelling, understanding and design of complex chemical and biological systems. In the past five years, his research has evolved to encompass biomaterials and regenerative medicine.

He currently leads two research projects; one involved in understanding regulatory processes in stem cells, and how they trigger fate decisions from a complexity perspective; the other is using more traditional molecular modelling methods to design small molecule mimics of cytokines that influence stem cell fate. These projects involve collaboration with the Australian Stem Cell Centre and other international stem cell groups.

Dave is a Fellow and past Chairman of the Board of the Royal Australian Chemical Institute, an Adjunct Professor at Monash University, and is the current President of Asian Federation for Medicinal Chemistry. He has published almost 150 scientific papers, confidential reports, and patents, co-edited three multi-author books, and transferred the Bayesian neural network technology to Bio-RAD as a software product, MolSAR. He a member of the editorial advisory boards of the journals ChemMedChem and Journal of Molecular Graphics and Modelling.

Author's Index

Index

Index